THE WEATHERING
AND PERFORMANCE OF
BUILDING MATERIALS

THE WEATHERING AND PERFORMANCE OF BUILDING MATERIALS

Edited by
JOHN W. SIMPSON BSc MSc AInstP
and PETER J. HORROBIN MA ARIC

Lecturers in the Department of Building of the University of Manchester Institute of Science and Technology

MTP
MEDICAL AND TECHNICAL PUBLISHING CO LTD
1970

Published by
Medical and Technical Publishing Co Ltd
Chiltern House, Oxford Road
Aylesbury, Bucks

Copyright © 1970 by John W. Simpson and
Peter J. Horrobin

SBN 85200016 2

PRINTED IN GREAT BRITAIN BY
BILLING AND SONS LIMITED
GUILDFORD AND LONDON

PREFACE

In the spring of 1970 we had the privilege of organizing a short course entitled 'The Performance of the External Surfaces of Buildings' at the University of Manchester Institute of Science and Technology. The course, which was closely related to research work in the Department of Building, was designed for architects, surveyors, engineers, and all those concerned with the weathering, durability, and maintenance of building materials. During the months of preparation which preceded the course it became increasingly obvious that the subject of the weathering and performance of building materials was both ill-defined and poorly documented. Consequently we felt the time was ripe for the publication of a new book devoted exclusively to a subject which in the past has usually (and unfortunately) been relegated to a minor subsection in standard texts on building materials.

Accordingly a specialist team of authors (most of whom lectured to the course) was enlisted to write about the principal materials encountered on the exterior of buildings—concrete, clay products, timber, metals, and plastics. These five chapters are complemented by an introductory chapter outlining the most important factors which contribute to weathering and performance.

It is our hope that the contents of this book will be of considerable help to both architects who design and detail a building and to the surveyors, engineers, and contractors who are responsible for its erection and maintenance.

JOHN SIMPSON
PETER HORROBIN

Manchester, September 1970

ACKNOWLEDGEMENTS

The editors would like to gratefully acknowledge the help and advice given to them on numerous occasions by their colleagues at the University of Manchester Institute of Science and Technology.

In addition, the editors are grateful to the many organizations who have freely allowed photographs, tables, and illustrations to be used in this book; full acknowledgement is given on the appropriate pages.

CONTENTS

	PREFACE	*page* v
Chapter 1	WEATHERING AND PERFORMANCE Brian Atkinson, BSc Tech MSc PhD Dip Arch ARIBA	1
Chapter 2	CONCRETE J. Gilchrist Wilson, FRIBA	41
Chapter 3	CLAY PRODUCTS H. W. H. West, BSc FGS FI Ceram	105
Chapter 4	TIMBER Gavin S. Hall, BSc MF DFor AIWSc	135
Chapter 5	METALS W. D. Hoff, BSc PhD AInstP	185
Chapter 6	PLASTICS Kenneth A. Scott, BSc ARIC FPI	231
	INDEX	277

Chapter I WEATHERING AND PERFORMANCE

BRIAN ATKINSON

BSc Tech MSc PhD DipArch ARIBA
*Lecturer, Department of Architecture and Planning,
The Queen's University of Belfast*

INTRODUCTION

Visual and compositional changes which take place in and and on the external surfaces of buildings are usually loosely referred to as weathering; but more recently these changes have been considered rather as part of the total performance of a building.

The term weathering implies the action of the weather: indeed, an older interpretation of weathering brings to mind mellowed brickwork, lichen-covered stonework, and silver-grey cedar boarding; but the situation is far from one involving the simple patinous action of the natural climate on external surfaces. There are many other factors which affect the performance of external building materials, and all the changes which take place are to some extent deleterious rather than beneficial to the fabric of the building.

The greater part of our stock of buildings exists in what could be defined as urban areas. Here, the natural climate is modified by the products of combustion and manufacture, by urban topography, and by the change, on a vast scale, in the nature of surfaces—from soil and vegetation, to asphalt, brick, and concrete.

Within the conurbations, mellowing effects are obscured by unsightly staining and blackening; organic growths are eliminated or retarded by the action of sulphur compounds; there is a marked absence of timber decay and insect attack compared with that found in rural areas; the breakdown of paintwork on timber substrates is unaffected; but the corrosion of exposed metalwork is generally accelerated.

All materials deteriorate. External building materials and assemblies are no exception to this law. Scientific observation shows that some deteriorate at a faster rate than others depending on a set of controlling conditions or causal factors. Quite often these conditions can be modified to alter the rate of deterioration to a preferred or acceptable level. The skill of a good designer lies in his ability to obtain an optimum performance using appropriate materials for a given situation.

Some authorities refer to weathering as the action of both the natural climate and atmospheric pollution, together with any other extraneous influences. With some justification the term 'weathering' could in this case be regarded as synonymous with that of 'deterioration', and the term 'climate' replaced by 'environment'; but it is also recognized that visual enhancement of a building may occur simultaneously with the compositional breakdown of the underlying materials.

The total performance of a building refers to the ability of both internal and external materials to fulfill their intended function over the useful life of the building. It also refers to the logic and efficiency of spatial enclosure and internal environment during use, as well as to the maintenance of satisfactory appearance. The performance of external materials forms only one part of the total material performance of a building.

For the external fabric, a logical order of consideration should begin with an understanding of the behaviour of the major materials; follow with an appreciation of the design elements of which they form part; and move to the complete wall or roof, considering for example, sectional profile or surface geometry which largely influence water-shedding characteristics, surface temperature, and wind flow. The interactions and relationships between the external parts of a building are equally important as the parts themselves. The behaviour of an external surface is inseparable from the behaviour of the complete external section, and in many instances is directly affected by the function of the building and by the habits and activities of the occupants.

The performance of the external fabric has special significance at the present time in terms of maintenance costs. Over the last decade or so the costs of external maintenance, as evidenced in local authority records, have risen at a much greater rate than internal or other costs. For Manchester, one of the largest local authorities, the factored external maintenance costs per dwelling exceeded those of internal maintenance in 1952, and exceeded the sum of all other maintenance costs in 1967. Internal maintenance costs actually fell between 1950 and 1967. This authority has a total stock of over 80 000 dwellings and shops. If economic priorities are to be established in the domestic field the performance of external materials and assemblies should have first consideration. For commerical enterprises, ordinary cleaning costs can assume even greater importance than the cost of repairs and renewals to the fabric of the building.

The differences between external and internal performance are essentially bound up with differences between the effects of wear by use and that of the environment. User wear does not extend far beyond external openings, communal balconies, or staircases. Externally there is little scope for the use of new and less durable materials because of climatic extremes. The internal environment generally allows a greater freedom of choice and experiment because of its inherent stability.

The measurement of the performance of external materials presents difficulties. The problem can be partly resolved by reflecting that because all materials deteriorate there is always a loss of performance with respect to time. Changes in appearance and composition always take place, and a measure of the rate of deterioration can provide a measure of minimum performance. Some authorities believe that the cost-in-use method of assessment is a valid measure of total performance, but this approach is open to question because it ignores amenity values, changes in appearance, and is based on maintenance costs rather than on direct observations of buildings. Nevertheless, an analysis of maintenance costs does provide a partial measure of total material performance; it is the only quantitative measure of its kind at present available to the industry.

THE NATURE OF THE PROBLEMS

The design of the external fabric of buildings is still largely based on concepts embodied in traditional building practice as distinct from those based on an understanding of the physical laws governing the fundamental behaviour of materials and components.

There is historical justification for maintaining this situation, for buildings taken as a group have always been one of the most valuable and durable of the nation's fixed assets (1, 2), individually requiring relatively little attention after erection.

The acceptance of changes in design has been generally slow; slow enough to regard them as evolutionary. The useful life of most buildings gives ample time for a particular method of construction, or the use of a new external material, to be justified or condemned. The inexorable test of time has filled the building graveyards with condemned building systems and external materials; only a handful have survived. The significance of the evolutionary process is exemplified in the current efforts to introduce industrialized building methods. White (3), for example, has described the pre-1939 and post-war failures in this field; the third stage of fairly widespread acceptance has apparently been reached, but the final verdict, fifty years hence, has still to be pronounced.

Experience with new building types has shown the need for considerable caution because one of the commonplace symptons of changes in technique and in the use of new materials is their breakdown or failure. This is particularly true of the external fabric.

The most perfunctory examination of buildings erected during the last two decades in any town or city, will reveal that in many instances the reasons for certain external details used prior to this period are not widely understood. The neglect of even elementary safeguards in many designs has resulted in premature deterioration in appearance and composition.

Articles are sometimes published drawing attention to this kind of problem (4–26), but whilst the more serious examples are brought to the notice of the professions concerned, few studies by case-history or of performance have emerged which indicate, on a priority basis, the day-to-day forms of building failure which are most likely to occur with particular forms of construction in a given environment.

Studies abroad reveal a similar state of affairs. Jaeggin and Brass (27), after a search through the available literature, reaffirm that an activity that appears to be conspicuously missing from the building process is that of post-construction study—the process which enables designers, builders, and owners to learn of satisfactory and unsatisfactory features of their buildings by examining them after a period of use.

Legget and Hutcheon (28), refer to the current American interest in the performance concept as applied to buildings. They advocate performance tests for materials and assemblies based on field experience rather than on standard tests.

Page (29), referring specifically to the effects of climate on design, which is one of the most important influencing factors affecting external performance, also advocates study by case-history. He maintained that the theoretical situation does not exist, which is probably true.

At the 1965 RIBA conference, *'The maintenance of buildings'*, E. D. Mills (30) repeated a plea made ten years earlier by W. Allen, in a similar conference at Torquay, for a feedback of information on the behaviour of techniques, details, materials, and ideas that had proved successful. There have been many pleas of similar nature asking for the establishment of an unbiased feedback of information on both successes and failures of building designs; the feedback being directed at the designer.

Where changes in materials and methods of construction take place, additional problems also arise during erection, for the builder or contractor frequently relies almost entirely on traditional trades and methods of construction. The traditional rapport between designer and builder is extremely sensitive to such changes.

Despite the advent of many new materials in the building industry, and the early post-war (3, 34–6) and recent use of radical methods of construction (37–45), particularly in the field of the dwelling-house, the bulk of external materials used during this period, taken for definition purposes up to 1965, still consisted of brick, timber, concrete, and glass with rather lesser exposed areas of steel and aluminium alloys. Externally the most prominent changes are not to be found in the use of new materials but in the assembly of those already well known and in the varying proportions that each are used.

Since 1950 two basic changes have been accepted in external construction. The first is the large-scale introduction of various systems of light and heavy cladding, and infill units, and the second is the increased use of the flat roof. A great deal of trouble has been experienced in this country and abroad with both these forms of construction.

The international conference at Rotterdam in 1959 (46) focused the many and widespread problems of the flat roof. Schaupp (47) has emphasized that nearly all the troubles derive from faulty assembly rather than unsuitable materials (48–50). Troubles with this type of roof in Great Britain have been serious enough to merit special investigations by local and county authorities (51, 52). In 1960, Howitt (53), the City Architect to Manchester, was pressed by the Buildings Sub-committee of the Education Committee to justify the continued use of flat roofs for schools, in view of the number of failures and the consequent repair and maintenance costs. In his second report to the sub-committee the initial planning and economic advantages of this type of construction were stressed. The County Architect for Hertfordshire produced a similar report for his education committee in 1958 (52).

It would also appear today that the system of building construction which involves the use of a structural background covered with a non-structural, but self-supporting weatherproof envelope, is here to stay for some considerable time. There are several systems of enclosing a structural background which lie somewhere between the traditional solid wall and the curtain wall. In all these methods at least part is prefabricated off the site. The essential characteristic is that they are not structural and are normally assembled out of unit panels fitted directly together, or onto some kind of frame, which is, in turn, carried on the structural background. Øivind Birkeland (54) reminds us that they are typically unconventional walls and they present many problems which must be solved anew. In 1959 Rostron (31, 55, 56) in a survey of 67 light-cladding constructions, found that 42 per cent were defective. Movements found in light-cladding have been found to permanently damage panels and double-glazing units, particularly those with a glued or yielding seal; most of the problems arise at the joints. In 1960, a field study by Bastiansen (57) in Norway, revealed that 10 per cent of 1300 double-glazing units were defective.

In 1962, Atkinson (58) found several fairly new buildings of light-cladding design in the Manchester area which allowed water to penetrate the joints, and promote the deterioration of parts of the cladding. The contributory causes of failure were considered to be poor sealant formulation, badly designed joints, loose specification, and incompetent handling and application of jointing materials. Dry joints were found to be the most successful. Both Hunt (59) and Birkeland (54) have outlined the many defects inherent in this form of construction (60–2).

In the vast majority of buildings today, economics are frequently the principle determinant of design and hence of durability. Initial costs are the most immediate concern of the owner; annual costs usually include interest on the capital outlay and sometimes sinking fund payments; maintenance charges are necessary to keep the asset intact; and the operating costs must also be included in total costs.

The taxation policy of this country, however, favours low capital outlay for certain types of building, thus encouraging the use of cheaper materials requiring more maintenance. Marr (63) has lucidly outlined the present position. He shows, for example, that for a profit-making concern expenditure of £1000 on maintenance actually costs the concern only £463, though it consumes £1000 worth of the nation's resources.

On one hand, for example, we may condemn the use of asbestos cement products externally in urban areas because of their propensity towards rapid disfigurement, embrittlement, or carbonation cracking; but, on the other hand, we eagerly accept them on economic grounds. The problems which frequently lack attention are not those concerned with the detailed intrinsic properties of materials, but those due to indirect extrinsic causal factors which encourage deterioration in other ways.

The magnitude and intangible nature of the total problem cannot be underestimated. For instance, various statements given by a number of authorities estimate that between 30 and 50 per cent of the labour force of the building industry is engaged on maintenance work. Ministry of Works figures for 1956 show that about 30 per cent of labour is involved (64); in 1965 Parker (65) gives 40 per cent; and Building Research Station estimates indicate that the probable figure is nearer to 50 per cent (66).

Despite the frequent questioning of these figures it can be shown mathematically that values of this order are to be expected. For a fixed labour force producing building units of a given service life there is always a limit to the total stock and this limit or saturation is approached according to an inverse exponential function; at the same time, the labour force available for new building work progressively diminishes as the stock increases because of the need for maintenance, thus reducing the output still further. Modifications to these simplified assumptions give a more realistic evaluation, but the conclusions are much the same. At present therefore, half the total labour force of the building industry maintains a total building stock which, following Redfern (1, 2) can be valued at about £50 000 million (1965 values). The cost of this maintenance work is estimated as £1100 million (1965 values), or approximately 3 per cent of the current value of our gross fixed capital investment in building works (67). This is remarkably good value; so much so that the introduction, for example, of a new external material which can compete successfully with those already in use, and be fully accepted, is a rare event.

Maintenance costs are clearly useful as a guide to the types of external deterioration that arise, but the method of recording these costs is rarely in sufficient detail or in an appropriate form to be readily useful to the designer. We know, for example, that for local authority dwelling-houses external maintenance costs account for about 50 per cent of total costs; and external painting represents about 75 per cent of external costs, or 32 per cent of total costs. The properties of paint are fairly well documented and the effects of moisture, ultra-violet radiation, and oxygen, which are the principal agents of deterioration, have been determined; but there is a marked absence of activities which consolidate the work of the economist and chemist by examining the real situation on site. There are, for example, no surveys of paint failures on buildings which differentiate between the effects of pollution, detailing, orientation, and substrates, using readily available techniques provided by the statistician. There would appear to be a gap in the field of building research which the designer is hesitant to enter because it appears too scientific; and which, for example, the chemist and physicist ignore because it is not scientific enough.

Definition of terms

Durability is a measure of the rate of deterioration. A definition in most dictionaries (68, 69) defines durable as 'able to endure', or 'lasting', or 'resisting wear'. The BSI Code of Practice (70) defines durability as the quality of maintaining a satisfactory appearance and satisfactory performance of required functions. Rates of deterioration or durability are measured in the Code in terms of the minimum years of satisfactory life and are graded into three qualities; a better quality material or design element has the property of greatest durability (Grade A), or minimum rate of deterioration. The *Report of the Committee on Building Legislation in Scotland* (71) gives a most appropriate definition of durability. This states that from the point of view of the building as a whole, durability is the product of a large number of factors—the type and quality of materials chosen; the fitness of the design for the use to which the building is put; the stability of the site on which it is erected; the degree of exposure to which it is subjected, and other factors. Durability is not an isolated function, it is the product of many.

Deterioration is almost the antonym of durability; but durability is essentially a 'rate' whereas deterioration is a 'state'. The Oxford dictionary definition of deterioration outlines the word as meaning a process of becoming or making worse. The word 'faulty' is implicit amongst others in the word 'worse'. Greathouse and Wessel (72) state that the deterioration of most materials is associated with a transition from a higher energy level to a lower energy level. Schaffer (73) defines the related term 'weathering', as a series of processes which bring about unfavourable or favourable changes in the appearance or condition of materials. The deterioration of building materials is usually considered to be the result of their unfavourable weathering. The durability or rate of deterioration can only be determined by observation.

The deterioration of buildings, as defined here, refers to the loss of value of some or all of the materials out of which they are built and to a decrease in the ability of any number of the products or elements involved to fulfil the functions for which they were originally intended.

A product in the building industry may range from a screw to a complete assembly. The forms of external deterioration that arise may be visible or obscured beneath the outer fabric; they may be aesthetic or functional and thus be open to subjective or objective assessment, or both. The definition of deterioration which includes the total building must include parts inside and outside as well as above and below ground. The failure of services and equipment is excluded from the definition, but may be the cause of deterioration internally or externally in certain cases.

It is convenient to distinguish between deterioration in appearance and deterioration in composition, though they often arise for much the same reason. In the case of the former it frequently happens that no attempt is made to rectify the changes because they may be held to be favourable, or because neither the physical properties of the fabric nor the functions within the building are affected. In certain instances however, aesthetic deterioration can result in a building giving an apparent decreased service or value. For example, it is well known that for many city commercial buildings a prestige value is attached to the maintenance of pristine external surfaces. In these instances it is possible to calculate an equivalent annual value, or capitalized cost, for cleaning costs which can then be added to the total maintenance costs.

Deterioration in appearance is associated with tonal and colour changes of the external surfaces, either even or variegated, lighter or darker. The physical degradation or erosion of surfaces can cause changes in texture, as well as changes in shape or visual definition. The original design conception, its form and massing, scale and proportions, together with its psychological impact and visual interest, can easily be destroyed in this way. Widespread effects on an urban scale can be extremely depressing not only to the trained observer but also to the layman, even if the reasons are not consciously understood.

The breakdown of a material in a truly 'material' sense is embraced by the term 'compositional failure'. Functional requirements, such as making good compositional failures, are usually given first priority where buildings are concerned. This is reflected in the number and types of jobs which appear in maintenance records. If a material fails to do as it is intended then the internal functions of the building are liable to be disrupted. It is possible to define compositional failures in a building as all those physical, chemical, and biological changes in the materials out of which it is built, and in the larger elements built up from these, that under the joint actions of use and environment decrease its ability to fulfil its original role.

ISOLATION OF THE CAUSAL FACTORS

The process of visual and compositional deterioration which takes place in the fabric of a building is one which, by its very complexity, defies analysis as a whole. Observations indicate that there are many interacting variables which individually or jointly can significantly affect the rate of deterioration of a material or design element. Surface form or geometry, orientation, detailing, and exposure are, for example, a few of the less numerically determinate variables. Others such as rainfall, solar radiation, wind speed and its directional frequency can be more closely defined.

It has been found that if materials are chosen for a given situation on the basis of individual soundness then this does not necessarily imply equivalent performance when used in combination. For the external fabric, material combinations are as important as the materials themselves.

In our loose everyday thinking about the problem, we seize on certain features which excite the imagination, and build around these features a framework which is often little more than a caricature of limited personal experience. Nevertheless, if an analytical approach to the study of building performance is to be made, some attempt at establishing an operational framework is essential. For the problems of external or internal material performance it is convenient to base such a framework on causal relationships, to separate the intrinsic and extrinsic material variables, and to investigate the dependent total measure of durability.

A marked feature of performance studies that have been

undertaken in the past is the absence of any kind of information which will enable the relative effects of the controlling factors, however defined, to be assessed. The prime reason for this is that in most cases such information has to be obtained first-hand, and this would invariably take longer than an examination of the buildings themselves. Field studies (74–91), for example, of the urban topoclimate are still very limited in approach and have not yet captured the interest of designers who have opted to undertake research work. Much of this kind of work has been undertaken by geographers, meteorologists, and aeronautical engineers who are often unable to establish priorities, or apply the results to the needs of the building industry, because of lack of a suitable directive and personal building experience.

A number of authorities have summarized the problem of deterioration by grouping the causal factors in various ways. These are diverse in approach, probably because there is some confusion between the grouping of factors and the grouping of individual variables, as well as failure to distinguish cause from effect.

The BS Code of Practice (70) on durability refers to a number of 'deteriorating factors' which include: atmospheric and climatic action; wetting and drying effects; soil and ground water action; rodent, insect, bacterial fungal, and plant action; water supply; electrolytic action; contact or association of incompatible materials; specific chemical action or chemical changes in materials; wear; impact and vibration; action of cleaning and cleansing agents; action of domestic or industrial wastes; and accidental causes, including fires, lightning, and floods. There is considerable ambiguity in this collection of factors, and some confusion between the nature of the applied forces and the corresponding reactions; or more simply, between causes and effects.

Greathouse and Wessel (72) draw a distinction between material reactions and applied forces by considering initially what are defined as the 'agents' of deterioration. These are named as the direct and indirect climatic agents, physical and chemical agents, and biological agents. In their work there is a broad overlap, both in the description and definition, of agents referred to under the heading of 'climate', and between those appearing under the other two headings.

Mills (30) has listed what he considers to be the most important factors contributing to the deterioration of buildings, and these include moisture, natural weathering, corrosion and chemical action, structural and thermal movement, and user wear and tear. In addition, 'detailing' and 'material' performance are mentioned separately. A notable omission is the 'standard of workmanship'.

H. E. Buteux (98) has outlined the relationship between design and maintenance. He maintained that deterioration and movement in buildings are the result of moisture, natural weathering, atmospheric pollution, material deterioration, fungi and insect attack, corrosion, chemical action, structural movement, and user wear and tear. Workmanship, detailing, and specification are also mentioned. These factors are similar to those named by Mills. The isolation of corrosion from chemical action, and moisture from natural weathering, as well as the inclusion of material deterioration is illogical if the usual definitions of terms are assumed.

Chaplin (99) in a study of army dwellings has presented an analysis of maintenance costs and summarizes the causes of maintenance under the headings of fair wear and tear, frost damage, faulty materials or workmanship, design or specification faults, miscellaneous repairs, unnecessary work, and chargeable user damage. If the causes of maintenance are synonymous with those of deterioration, as Chaplin assumes, the mixture of the particular with the general is perhaps admissible.

An interesting and novel approach at defining the rate of deterioration of a material has been made by Brooks who derived an equation which gave a numerical rating for a given location or site (100, 101).

Reiners (102), in a study of the maintenance costs of local authority housing, makes reference to a number of more fundamental factors which influence the incidence of maintenance work. These are named as exposure, workmanship, quality of materials and design, and details. In addition, the factor of organization is included, but to whom this factor refers is not made clear. The factor of 'wear by use', included by some authorities, is omitted.

It is highly likely that such a diversity of interpretation reflects the professional interests of those concerned. However, it is possible to discern a similarity of approach between most of these which leads to a fairly simple grouping of the causal factors.

Greathouse, Mills, Buteux, Chaplin, Brooks, and Reiners, for example, grouped the causal elements in different ways according to their own experience, but recent performance studies (103) indicate that four basic causal factors can be isolated which apply equally well to the behaviour of either internal or external materials.

The factor of *environment* (E) is common to most interpretations of the problem. Indeed, it is the sole criterion of Brook's approach. Environment comprises a set of elements such as air temperature, rainfall rate, wind speed, sulphur dioxide level, and smoke concentration which are well defined; and also a number of ill-defined elements such as topography and soil conditions. These can be picked out of the list given in the BS Code of Practice (70). The environmental factor has special significance in respect of the external fabric, it is the one over which the least control can be exercised, and in practice has the widest range of effects on external materials and design elements.

Another causal factor is that of *use* (U). Chaplin (99) was especially concerned with 'wear by use' because of his experience of excessive damage in army houses, and he differer tiates between 'fair wear and tear' and chargeable user damage. Deterioration by use depends on the behaviour and habits of the occupants which are generally reflected in the function of a building or part of a building. There is a considerable degree of interaction between the factor of use and that of *design* (D). The designer must take full account of the effects of use in his selection of materials, planning, and manner of detailing. For the external fabric the factor of use is of marginal importance.

The design of a building embraces all the technical, spatial, and aesthetic skills of a designer as applied to the owner's requirements and budget. The factor of design includes a number of causal elements which are beyond numerical interpretation. The selection of materials and their specification are determined by the designer mainly on a basis of experience and partly by training. A designer should be aware of the performance of the design details and elements that he specifies, and of the materials out of which they are made, in order to present a rational cost-versus-performance case to the building owner. A design failure, therefore, can sometimes be attributed to the policy of a building owner who demands the lowest initial outlay at the expense of performance. Mills (30) and Buteux (98) refer to the contributory factor of 'detailing; Chaplin (99) refers to 'design and specification faults; and Reiners (102) also includes 'design detailing' as a fundamental causal factor. The design factor excludes intrinsic material properties but includes the manner in which they are assembled.

Another causal factor which can be isolated is that of *workmanship* (W). Workmanship refers to causal elements which cover the total effort of the builder in erecting a building. It also includes his responsibility for using sound materials as described or implied in the specification, drawings, or by established practice. Poor workmanship which results in the deterioration of a building is brought about by a combination of negligence and incompetence generated by indifference to the job in hand or insufficient skill. Standards of workmanship do, of course, vary within the framework of any design and undocumented deviations from drawings and specifications nearly always occur, often with no detrimental effects, and sometimes to the benefit of the building owner. Within the building trade there are 'expected' standards of workmanship which are closely related to the function of the building. Working drawings for many traditional buildings are often little more than a symbolic representation of what the builder has to undertake. A flexibility of approach is thus adopted by designer and builder with limits implicit in the kind of job in hand. The switch from this method of working to the more rigorous approach of industrialized building requires changes of attitude as well as changes in methods of assembly. Reiners and Chaplin both refer to workmanship as an important factor; Chaplin separates faulty materials from poor workmanship, which is perhaps justifiable in view of the fact that the manufacture of a great number of building materials is outside the control of the builder. Nevertheless, the onus is upon the builder to use sound materials. In the present discussion the term workmanship is intended to cover faulty materials as well as workmanship in the craft sense.

A measure of the total deterioration may be symbolically equated with the causal factors mentioned above and with the intrinsic properties (P) of the materials involved.

$$\text{Total Deterioration} = f[(D), (W), (E), (U)][(P)]$$

In the field all factors are operative though for external materials the influence of use is minimal. Any total measure of deterioration depends more on the interaction between factors than it does on factors considered in isolation.

It is important to distinguish between the effects which take place on or in a material because of its intrinsic properties, and the extrinsic or causal factors which arise independently to create these effects. Causal factors and material properties are related, and either of these entities, however expressed, may be used as an independent measure of the rate of deterioration for a situation described by the other.

For the external fabric only the basic intrinsic properties of a few major materials need to be understood, on the other hand the influence of the causal factors is infinitely conplex. It would appear that any attempt at presenting the results of performance studies for practical use can only, at best, be selective. For the designer a most convenient scale of presentation is by design element. He need not pursue studies of material properties with as much rigour as is sometimes considered necessary because there is far more to be gained from studies of the behaviour of materials in combination at design element scale and larger. Evidence indicates that this is largely true if performance is a criterion of good design. Traditional studies of external construction tend to present design details with only cursory discussion of performance and causal factors. Studies of materials have also tended, in the past, to be largely lists of properties combined with methods of extraction, treatment, or manufacture.

The causal factor of design is related to events which take place mainly before a building is legally complete. Maintenance work is normally carried out throughout the life of a building, and for this reason the factor of workmanship extends over a longer period. The factors of environment and use are related to events occurring after completion, and for the external fabric the factor of environment is the most important of these.

SITE OBSERVATIONS

It was stated earlier that the most prominent changes in external construction are not to be found in the use of materials but in their assembly, in the varying proportions in which they are used, and in the detailing. In most of our buildings neither brick, timber, metal, or glass are replaced by new materials, only the proportions of each that are used are varied. For example, a wide range of developments in wall design can be found. The use of a combined external door, in the form of an infill unit is a type of design common to the period. A similar development can be observed in window design. The simple window unit is often replaced by combined multiple units which incorporate opaque insulated panels. The ultimate development of the trend towards larger areas of infill is the curtain wall, which finds expression mainly in the façades of commercial buildings.

The basic principles of structural frame, box (or shell), and load-bearing wall, remain largely unchallenged. Changes in the proportions of the common external materials are sometimes accompanied by internal changes in structural form. At some stage in the design it would appear to be a economically viable to replace the simple load-bearing wall by a more easily erected structural frame because the part played by the former load-bearing material in enveloping and protecting the interior, becomes redundant. The point at which either the dictates of an owner, or the economies of the problem assert themselves is difficult to define. The designer does not always act to make good use of economic advantages that may result from such a change.

From the time of Paxton's venture of 1851 all the variations of solid-to-infill ratio have been in use. The pressures of arbitrary design choice, and even fashionable forms have always been greater than a choice governed by the marginal economic advantages which differentiate methods of construction that are available at any one time. Site observations show that both large and small decisions are made independent of cost implications. The idiosyncracies of a prospective owner, those of the occupier, and the function of the building obviously affect such decisions.

A measure of the influence of the factor of environment on the deterioration of materials is reflected in the small number of new major materials that have been introduced externally compared with those introduced internally. During the last two decades one major internal material, or group of materials has successfully emerged—that of plastics. There have been no comparable developments in the field of external materials, but so great has been the impact of this versatile material that it can be predicted that it is only a matter of time before it is developed further and integrated into external design as a major material. The external durability of this material has been proven already by its development and use in the form of such items as guttering, soil stacks, rainwater pipes, and a host of other minor external components.

Performance studies that have been carried out in this country so far (103) show that a great number of the problems of deterioration are either traditional, or extensions of traditional problems, which are already well documented in the professional literature. Such a situation has arisen because relatively few materials are in evidence. Common defects such as the gradual breakdown of mortar work, the peeling of paintwork on timber windows, or the disintegration of putty runs are well known. It has been found that increases in the proportion of undivided glazed areas or impervious surfaces lead, for example, to increases in rainfall run-off which can aggravate many of the associated traditional problems; resulting for instance, in the exacerbation of paintwork failures on external sills, increased rates of putty or mastic failure on lower members, and the saturation of adjacent brickwork at sill ends. An even more untenable, but typically traditional urban problem is the disfigurement of buildings by pollution and dust. Only the most obdurate designer could believe that the use of large areas of exposed concrete would render his work less susceptible to disfigurement in this way than the works of his forefathers who had to make use of sandstone and limestone. The most expensive problems are generally associated with new forms of construction. Those of light-cladding and flat roofs, for example, have been mentioned already.

Failures in composition and appearance have been defined and the terms 'functional' and 'aesthetic deterioration' are often used to describe the same respective effects. The continued disintegration of an external material or its compositional breakdown may result in wider, more serious problems in that part of the building because of reduced structural stability or weatherproofing properties. For example, the continued lack of paint on exposed timberwork, particularly at the joints, may necessitate complete renewal of part or the whole of the element concerned—a window or perhaps an infill unit. The continued spalling of a concrete beam due to, for example, a porous mix and perhaps hastened by poor detailing, would seriously affect the wall it supports if left unmaintained. For similar reasons the spalling of brickwork cannot be left untreated indefinitely. More often than not the breakdown of an individual material involves several other materials and the failure of a design element as a whole must be examined. Classification of compositional failures for analytical purposes is complicated by cause-and-effect relationships.

Independent of any method of analysis such as this the types of compositional failure that are observed in field investigations can be grouped into three broad categories; firstly, the constantly recurring traditional faults such as frost attack and paint failures; secondly, a minority group which includes the more exaggerated forms of traditional failure, such as movement cracks caused by the use of large concrete slabs, or leaks due to the failure of oil-bound sealants; and thirdly, another minority group, but of entirely new problems, such as excessive interstitial condensation in sandwich wall panels, or the effects of ultra-violet light on plastics. Most of the new or non-traditional problems are associated with radical departures from traditional construction rather than partial or transitional changes; but the extent of failure increases with the degree of departure, and appears to be a reflection on specification and design ability rather than on workmanship. Deterioration in the external appearance of urban buildings is largely dependent on two parameters—namely, atmospheric pollution and weather.

Site observations confirm that compositional failures can be considered in terms of individual materials on a priority basis; alternatively the external fabric could be analysed, perhaps more appropriately for the designer, as a series of design elements such as openings, roofs, balconies, etc., with accurate reference to causal factors.

INTERACTION AND SURFACE FORM

The necessity of considering the performance of combinations of individual materials, which are incorporated into well-recognized design elements, has been suggested. Extending this approach, it is possible to consider the interaction between groups of elements which form, for example, a complete wall or roof, and further, to consider combinations of roofs and walls and other parts which together comprise an entire external fabric. A balanced understanding of material performance should ideally cover all the successive steps of consideration leading from an individual material to a total building, which can rarely be type classified in its entirety. At some stage too the division between studies of internal and external materials disappears and judicious examination of internal external relationships is necessary. For example, the examination of the external wall as a single large element inevitably leads to an examination of its total construction and of the function and types of internal materials used in it. Observations indicate that the division between studies of building construction and building materials is quite meaningless when seen within the context of physical performance.

A great number of failures cannot be attributed to the behaviour of isolated materials, but are due to assembly effects associated with design elements. The effects of the agents of deterioration are to a large extent dependent on the logic of the designer as expressed in the construction detailing. In the case of a roof, for example, the design principle adopted for dealing with precipitation or interstitial condensation must be definite and without constructional compromise; the physical laws governing the behaviour of materials are inexorable and can never be ignored. The disintegration of flat, felt-covered roofs caused by the omission of a vapour barrier on the underside of the roofs is a common failure. All the materials used may be found to be inherently sound, but the combination of materials can be quite unsound—the result of illogical design.

Apart from construction based on correctly interpreted physical principles an over-all simplicity in element design should be directed towards easy construction, and should also facilitate replacement or repairs which may arise some years later. In most buildings differential deterioration inevitably leads to partial renewals or replacements. It would seem, on the basis of observations, that the most durable external parts should be the innermost components of a design element and structural members should be given preferential treatment. Many of the early forms of industrialized dwellings involved the use of light lattice steel structural frames which are generally impossible to repair or maintain without dismantling large sections of the external or internal wall surfaces.

An excessive number of light sections encourages corrosion at joints, and at the lower parts of building elements. Fewer, heavier sections perform better than large numbers of light sections. Experience with the post-war aluminium bungalow showed the dubious logic of a complete sandwich wall. These dwellings proved almost impossible to repair satisfactorily despite the short-term type of maintenance work required.

It is possible to judge the quality of the external fabric of a building design on the basis of the incidence of maintenance problems and the ease and economy with which repairs and renewals can be undertaken, but the factors governing whether maintenance is desirable are not necessarily related to the philosophy of minimum decay, and even the effects of a poorly conceived design detail can be obscured from the investigator if very durable materials are used.

All external materials deteriorate, and all parts of the external fabric deteriorate at different rates not only because of their composition but due to purely fortuitous location in or on the face of the building. Most of these differences, however, are marginal and do not merit special attention. Replacements and repairs often involve renewing materials which do not, by themselves, need renewing but for the sake of access, or because of the method of assembly, or simply because of the need to produce an acceptable workmanlike finish, they are replaced without further consideration.

It is clear from site observations that orientation, and surface geometry or sectional profile, are of paramount importance in determining the extent of failure in a given situation. The rate of deterioriation of a given material varies considerably according to its location, in the geometrical sense, in an external surface. This is particularly true where climatic agents of deterioration are concerned, such as rain, wind, or surface temperature. The ability of a wall, for example, to shed rainwater away from its surface depends wholly on the sectional profile in cases where the rate of rainfall exceeds the surface absorbtivity. It has been found that the rate of deterioration of mortarwork above damp-proof course level increases towards the higher parts of a wall, at the ends of sills away from the prevailing driving rain, towards corners, and sometimes behind pipes. Little is known about the exact distribution of rainwater over buildings, but the consistent patterns suggest that this type of failure is related primarily to the direction and distribution of regional driving-rain as modified by surface relief and profile rather than to frost attack or general attrition.

The rainfall run-off from the ends of sills often results in the premature deterioration of mortarwork in these areas. In other cases leaks behind rainwater pipes, or around balcony gullies, are a potential source of a trouble and add to the over-all or general pattern of mortarwork decay. The extreme exposure of chimney stacks, which is well recognized, results in early disintegration. The effects of temperature changes and differentials, sulphate attack, and severe exposure to driving-rain combine to accelerate normal mortarwork decay in such a location; experience has shown that brickwork chimneys built close to the ridge and in normal mortarwork (1 : 3 to 1 : 1 : 6) require attention after about twelve years. Wind tunnel research has shown that for a pitched roof the highest wind speeds occur over the ridge; at this point the streamlines close and a vertical wind component produces high rainfall incidence angles. Surface geometry has considerable effect on the incidence of driving-rain, but the nature of the surface itself is an additional factor affecting rainfall run-off.

The surface geometry of a façade can equally well provide protection for materials. For example, the cheapest paints will survive far longer than average if protected from sunlight and moisture. In this country (104) these places ccuro below eaves, along the tops of window heads, and in any sheltered recess. The effects of ultra-violet radiation are so well recognized that it constitutes one of the two major considerations (the other being moisture) in artificial weathering. Yamasaki (105), who tested glycerol trilinoleate films under varied conditions, rated oxygen and ultra-violet radiation as the most important agents of deterioration, and the combined effect of oxygen, ultra-violet light, and moisture as being only marginally less destructive. Protection from any of these agents has a marked effect on durability: surface profile and the proximity of adjacent parts are therefore important in this respect. Observation shows that external variations in deterioration, visually or materially, are rarely the result of careful deliberation but simply effects incidental to the functional behaviour of the design elements.

Visual changes which take place on any external surfaces in an urban area are due mainly to the interaction of atmospheric pollutants, wind, and rain. The form and extent of disfigurement is again basically dependent on surface profile and also on the surface charcteristics of the chosen material; texture and absorptivity being the most important. The variegated results of pollution staining are part of a familiar urban visual scene. Incidental to the traditional and usually essential need for a sill throating is the characteristic formation of what, here, is called the urban soot fringe. This is the common surface disfigurement found below most horizontal projections on parts shielded from direct rainfall or rainfall run-off. The area of wall disfigured in this way is normally determined by the vertical section; and differences in staining are caused by variations in the width of the projections, which shield the walls and divert rainfall run-off. Visual effects are usually considered to be unpredictable, but observations have revealed (103) some of the basic principles involved; such effects are ignored by most designers who design on the basis of an unblemished external fabric.

There has been a recent interest in the nature of the

microclimate about buildings. The influence on external deterioration of adjacent solid bodies such as neighbouring buildings and walls, and the influence of objects with some wind penetrability such as trees or hedges, at the distances and of the sizes common in residential areas, are only marginal when compared with more immediate surface effects. The prediction of the failure of a given external material depends on the ability to predict the relative effects of a succession of variables which begin with the nature of the material itself and end with more macroscopic considerations such as the influence of regional climate or user behaviour.

Deterioration in appearance and composition can be considered under four broad headings related to the principle design elements, and to other items such as external plumbing. A natural division appears to lie between the problems of the roof and those of the wall. Schaupp distinguishes between the separate functions of wall and roof (32, 33), and in this country Lea (106) has indicated the adoption of this approach by the Building Research Station and other bodies. The relationships between design elements as observed on site indicate that a suitable division may be made between the wall, openings, the roof, and other special features such as the design of balconies and canopies.

EXTERNAL ENVIRONMENT

The causal factor environment is of special significance in any study of external deterioration. It has already been stated that this factor is the one over which the least control can be exercised, and in practice has the widest range of effects on external materials and design elements.

If performance studies of the external fabric are to be based on surveys of groups of buildings, then to be of any value the results must be related, as far as possible, to measurements, or at least an appraisal, of the natural and man-made climates of the areas in which they are situated. The aim must be to relate survey findings to the relevant environmental variables, but at the same time it is necessary to avoid overweighting any discussion with, for example, excessive meteorological data.

It is also impossible to achieve a balanced assessment of the results of performance studies without having first obtained climatic data which has been measured at the correct scale. It is sometimes argued, for example, that microclimatic data is more important than regional data, but the question of determining the relevance of regional or local environmental measurements does not arise in practice. Measurements of climate and pollution are equally as important at both scales; the right kind of measurements must be made, appropriate to the form of deterioration under consideration.

It has been stated that the effects of regional climate or local pollution can be modified by the location of the material on the external surface, and by its relationship with other materials in terms of design detail. For some materials the critical environmental conditions bear little relation to the regional or local measurements; for these, microclimatic conditions are more relevant. For other materials the influence of the principle elements of weather and pollution are negligible at any scale of measurement. To be relevant, data on pollution or climate must also be at a scale appropriate to the type of material chosen, its surface location, and the design detail into which is incorporated.

The validity of using meteorological climatic data as an aid to the design process has been questioned in recent years (75). Research interest has moved more towards detailed studies at the topoclimatic (79–94) and microclimatic levels, though only a little of this work has immediate relevance to practical building problems.

The difficulties of attempting to correlate the well-documented regional weather statistics with forms of deterioration actually found in individual buildings are immediately apparent after examination of the weather reports.

Climate is normally measured macroscopically on a regional basis. The object of such observations is to record the average course or condition of the weather as exhibited in the means and departures from these means. Representative air mass characteristics are determined as free as possible from the microclimatic variations of the site at which measurements are being taken. The standardized and accurately recorded statistics are tied mainly to the needs of aviation and shipping. Those concerned with building problems are particularly interested in the same microclimatic variations which are carefully avoided in regional weather records.

Geiger (98) has defined the microclimate as the climate not more than 1 metre above the surface of the earth, but the surface of buildings must be considered as a special extension of the ground. Kratzner (80), Duckworth and Sandbourg (76), Parry (82), Chandler (83–5, 93), and others (76, 82, 89) have made a special study of rural–urban climatic differences. Certain elements of a regionally measured climate are modified in and above urban developments, certainly beyond 1 metre. Measurements of air temperature and humidity made across Manchester by Atkinson (103) using the traverse technique confirmed the existence of cold air floods in the northern valleys and thermal winds. Rural–urban average differences in temperature, amounting to nearly 10 °F (5·5 °C), were not so marked as those found by Chandler across London but this was probably due to insufficient trial runs. On one occasion a northern valley was found to be 20 °F (11 °C), cooler than the Ringway recorded air temperature.

In the same way that the macroclimate is modified by, for example, the topography of a region, the local or what

Scaetta (107) has defined as the mesoclimate is modified in a number of ways by local development and relief. Hawke (94), for example, who has examined extreme temperature ranges on a regional scale has also examined the characteristics of frost hollows (95) which are well-known local phenomena.

Variation in colour, texture, and form, the juxtaposition of design elements on individual buildings, and the actual grouping of buildings dictate not only the microclimatic conditions in the immediate vicinity, but also the behaviour of the building materials on and below the external surfaces. Parmelee (96) has, for example, shown this in his measurements of the irradiation of vertical and horizontal surfaces, and Stagg (97) has measured surface temperature as high as 200 °F (93 °C). The latter result bears little relation to regional air temperature. Early work into paint failure mechanisms often failed to take into account such high temperatures. For example, the extensive investigation by Elm (108), into the failure of paint coatings due to moisture, included only limited temperature control.

The Meteorological Office weather reports are available at a number of major centres in this country. The data are punched on 80-column cards, and independent processing work is often undertaken at these stations, yielding useful results. For the purposes of a recent study of the performance of external materials in the Manchester area (103) the writer utilized standard processed meteorological data to devise what has been named a rainfall source diagram (Fig. 1.1). This proved to be of value both in the fieldwork and subsequent analysis by providing a relative measure of the amounts of rain reaching surfaces and re-entrant angles from different directions. Generally, it has been found that because of the way in which weather data is recorded, a great deal of numerical work produces very limited data for the designer.

32   PERFORMANCE OF BUILDING MATERIALS

If rain falls on a surface in equal quantities from all directions the source diagram is a circle (shown dotted)

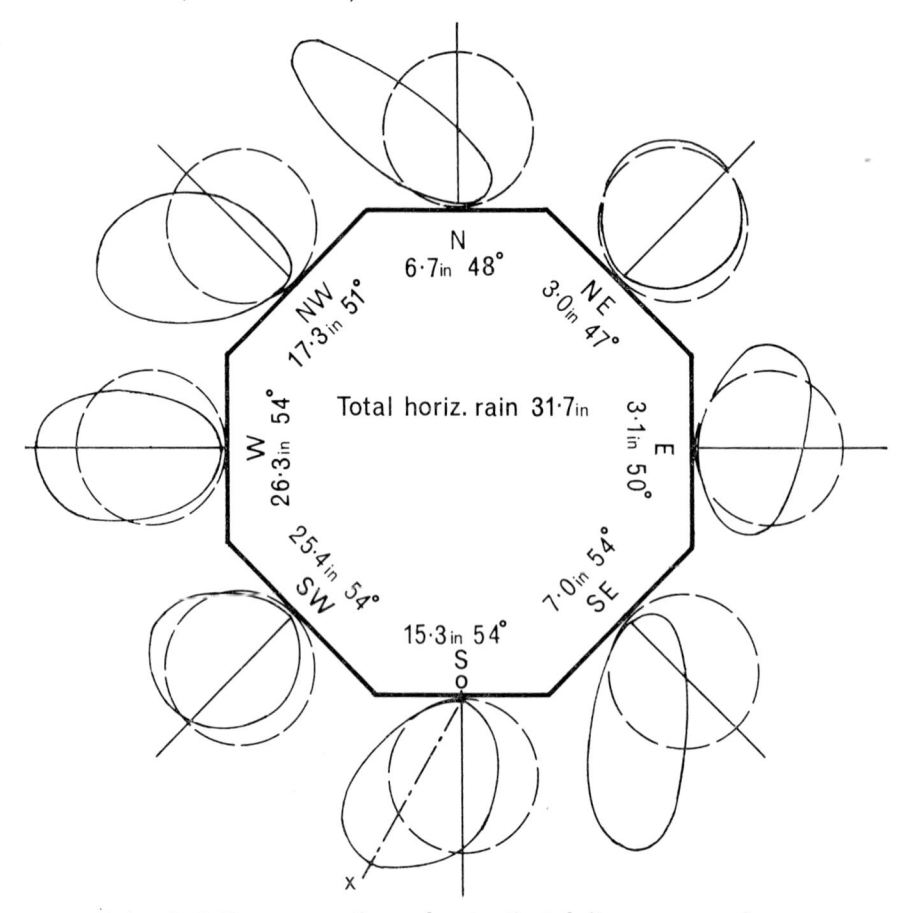

N 6·7in 48°
NE 3·0in 47°
E 3·1in 50°
SE 7·0in 54°
S 15·3in 54°
SW 25·4in 54°
W 26·3in 54°
NW 17·3in 51°

Total horiz. rain 31·7in

Note that the proportion of rain that falls on a surface from a given direction is equal to the length of the line drawn from the origin to the curve [e.g. For a south-facing vertical surface O–X represents the amount received from that direction]

Fig.1.1. The rainfall source diagram for Manchester.

Weathering and Performance

The principal elements of weather and climate are precipitation, temperature, humidity, wind, sunshine, air pressure, cloudiness, and fog. Not all these climatic elements are pertinent to external deterioration. The most important ones are precipitation, temperature, wind, relative humidity, and insolation. Relevant combinations are wind/rain, wind/pollution, and temperature/rain. Other combinations produce special types of failure in conjunction with other variables: the deterioration of organic materials for instance, due to the presence of moisture and followed by biological attack, must be mentioned.

Another important feature of the urban environment is atmospheric pollution. The effect of atmospheric contaminants on the external appearance of buildings is much in evidence on older buildings. Little is known at present on how to measure these effects or predict their severity in a given situation.

There is a considerable documentary evidence on the nature of atmospheric pollution but comparatively little, or none, on its relationship with visual changes in the appearance of buildings. The work of researchers such as Meetham (109–10), has indicated the types of pollutant, concentration levels (111), and probable urban distributions. The work of the Warren Spring Laboratories in England has provided data on the pollution levels in the major conurbations. This work is to be extended to cover other areas of the country. The influence of topography (112, 113), and climate (114–16) on atmospheric pollution has also been investigated. Sources of information on the extent of atmospheric pollution and on the regional weather are available in most large towns or cities. The appropriate statistics can generally be extracted from the records of the local Medical Officer for Health, and combined with data from the nearest Meteorological Office station.

Perhaps the most interesting feature of atmospheric

pollution which has emerged is the possibility that its effect on the deterioration of external materials has been considerably overrated. Standard test results, examining isolated variables, indicate that the type of atmospheric pollutants normally found in urban areas accelerate the breakdown of paint and the corrosion of ferrous metals (117–30). However, Oma (131) found that although a close correlation existed between corrosion and relative humidity (0·89), temperature (0·80), and precipitation (0·76), hardly any existed with atmospheric pollution (0·03–0·30) because of interaction effects. Ten environmental variables were measured simultaneously. Atkinson (103), in a paint failure count on timber windows in Manchester found that no significant differences in paintwork failures occurred between sites on the rural limits of the conurbation, and sites in the most polluted area of the city. In many respects the urban environment is less extreme than that found outside built-up areas, and a measure of protection is provided against the agents of deterioration.

Performance data based on real buildings in real situations are essential if a full understanding of the behaviour of external materials and design elements is to be achieved. This book deals with the performance of the four major groups of external materials in common use; namely, concrete, clay products, timber, and metals. The fifth group of external materials which is considered is plastics, which may be destined to become the most important of all groups.

REFERENCES

1. REDFERN, P. (1955). 'Nett investment in fixed assets in the United Kingdom, 1938–1953', *J. R. statist. Soc.*, Series A (General), **118**, part 2.
2. DEAN, G. (1964). 'The stock of fixed capital in the United Kingdom in 1961', ibid., Series A. (General), **127**, part 3.

3. WHITE, R. B. (1965). 'Prefabrication. A history of its development in Great Britain', *National Building Studies Special Report* 36 (London: HMSO).

4. DEVEREAU, M. (1962). 'Working details revisited 1. External non-loadbearing walls', *Architects' J.*, **136** (19), 1083-90.

5. —— (1962). 'Working details revisited 2. (a) Panel walls, (b) External claddings', ibid., **136** (20), 1137-9.

6. —— (1962). 'Working details revisited 4. Handrails and balustrades', ibid., **136** (22), 1241-6.

7. —— (1962). 'Working details revisited 5. Metal windows', ibid., **136** (23), 1289-94.

8. —— (1962). 'Working details revisited 6. Wood windows', ibid., **136** (23), 1295-1302.

9. —— (1962). 'Working details revisited 7. Swing and revolving doors', ibid., **136** (25), 1391-9.

10. —— (1962). 'Working details revisited 8. Sliding and folding doors', ibid., **136** (25), 1401-6.

11. —— (1962). 'Working details revisited 9. Rooflights and traps', ibid., **136** (26), 1449-57.

12. —— (1963). 'Working details revisited 10. Roof canopies', ibid., **137** (2), 99-104.

13. Anon. (1964). 'Building revisited. Hotel at Dover', ibid., **139** (23), 1275-80.

14. WATSON, D. I. (1960). 'Common building failures, 1: Granolithic paving', ibid., **132**, 214-6.

15. —— (1960). 'Common building failures, 2: Wall and ceiling plaster, cement/sand renderings', ibid., **132**, 335-6.

16. —— (1960). 'Common building failures, 3: The painting of steel, the painting of woodword', ibid., **132**, 471-3.

17. Anon. (1960). 'Comprehensive school', ibid., **132**, 535.

18. TURIN, D. (1964). 'The seamy side', ibid., **140** (13), 668-700.

19. THE SWEDISH STATE COMMITTEE FOR BUILDING RESEARCH (1959). *Flat roofs*, CIB Congress.

20. BUILDING RESEARCH STATION (1964). 'Design and appearance—1', *BRS Digest*, no. 45 (Second Series) (London: HMSO).

21. —— (1964) 'Design and appearance—2, ibid., no. 46 (Second Series) (London: HMSO).

22. HAMILTON, S. B., BAGENAL, H., and WHITE, R. B. (1965). 'A qualitative study of some buildings in the London area', *National Building Studies Special Report*, 33 (London: HMSO).

23. ADDLESON, L. (1965). 'Material for building—29: 3.00 water and its effects', *Architect and Building News*, **227** (22), 1033-7.

24. —— (1965). 'Materials for building—30: 3.00 water and its effects', ibid., **227** (23), 1095-7.

25. —— (1965). 'Materials for building—31: 3.00 water and its effects', ibid., **227** (24), 1141-4.

26. —— (1965) 'Materials for building—34: 3.00 water and its effects', ibid., **228** (2), 83–6.
27. JAEGGIN, K. W., and BRASS, A. E. (1967). 'A study of the performance of buildings', *Nat. Res. Council of Canada Tech. Paper*, no. 247 (Ottawa: Division of Building Research).
28. LEGGET, R. F., and HUTCHEON, N. B. (1967). 'Performance concept in building. Relation of testing and service performance', *ASTM STP* 423, p. 81.
29. PAGE, J. K. (1966). *Symposium on climate and design*. Welsh School of Architecture.
30. MILLS, E. D. (1965). 'The architect's approach to maintenance costs', *RIBA Conference* (London).
31. ROSTRON, M. (1959). *A study of light-cladding*, MA thesis, Liverpool University.
32. SCHAUPP, W. (1959). *The flat roof* (Nuremburg: Verlag Nurnberger Presse Druckhaus Nurnberg Cmbh and Co.).
33. —— (1967). *External walls* (London: Crosby Lockwood).
34. 'House construction' (1944). *Post-War Building Studies*, no. 1 (London: HMSO).
35. 'House construction second report' (1946). *Post-War Building Studies*, no. 23 (London: HMSO).
36. 'House construction third report' (1949). *Post-war Building Studies*, no. 25 (London: HMSO).
37. DIAMANT, R. M. E. (1964, 1965, 1966). *Industrialised building*, vols. 1, 2, and 3 (London: Iliffe).
38. HONEY, C. R. (1966). 'Industrialised building: Sponsorship and disciplines', *BRS Current Paper, Design Series*, no. 52 (London: HMSO).
39. ROSTRON, M. et al (1964). 'Pre-IBSAC Symposium', *Architects' J.*, **139** (26), 1413–36.
40. WACHSMANN, K. (1968). *The turning point of building: Structure and design* (New York: Reinhold).
41. RIBA (1967). *Industrialised housing and the architect: papers presented at a conference* (London).
42. RIBA (1965). *The Industrialisation of buildings: an appraisal of the present position and future trends* (London).
43. INSTITUTION OF STRUCTURAL ENGINEERS (1967). *Industrialised building and the structural engineer: a symposium* (London).
44. ROSTRON, M. and JORDAN, J. (1967). 'Forecasting of the future', *Science J.* (London).
45. BISHOP, D. 'Large panel construction: construction methods', *BRS Current Paper, Construction Series*, no. 9.
46. CIB Congress (1959). (Rotterdam.)
47. BUILDING RESEARCH STATION (1961). 'Flat roofs' *Library Communication* No. 1051 (translation of work by W. Schaupp).
48. —— (1964). 'Developments in roofing', *BRS Digest*, no. 51 (Second Series) (London: HMSO).

49. —— (1961). 'Built-up felt roofs', *BRS Digest*, no. 8 (Second Series) (London: HMSO).
50. —— (1949). 'The design of flat roofs in relation to thermal effects', *BRS Digest*, no. 12 (First Series) (London: HMSO).
51. HOWITT, L. C. (1958). 'Special report on the use of flat roofs on post-war schools', *Building Sub-committee of the Education Committee* (Manchester).
52. COUNTY ARCHITECTS' DEPARTMENT, HERTFORDSHIRE (1958). *Report on flat roofs* (London: HMSO).
53. HOWITT, L. C. (1960). 'Further report of flat roofs', *Building Sub-committee of the Education Committee* (Manchester).
54. BIRKELAND, Ø. (1962). *Curtain walls. Handbook 11B* (Oslo: Norwegian Building Research Institute).
55. ROSTRON, M. R. (1960). Series of articles in the *Architects' J.*, technical section, **131**, 337, 361, 404, 437, 518, 653, 724, 761, 966, and **132**, 115, 424, 540, 578.
56. —— (1964). *Light cladding of buildings* (London: Architectural Press).
57. BASTIANSEN, A. (1961). 'Factory-sealed double-glazed window units in Norwegian climate', *NBRS Reprint*, no. 53.
58. ATKINSON, B. (1962). 'A study of some sealants used in light-cladding and similar work', *Manchester Building Centre Prize 1961, Report* (Manchester).
59. HUNT, W. W. (1958). *The contemporary curtain wall, its design, fabrication and erection* (New York: Dodge).
60. BUILDING RESEARCH STATION (1957). 'Light cladding', *BRS Digest*, no. 98 (First Series) (London: HMSO).
61. —— (1957). 'Light cladding', *BRS Digest*, no. 99 (First Series) (London: HMSO).
62. AITKEN, D. F. (1967). 'Keeping the wet out of buildings', *Guardian Report*, 13 April (Manchester).
63. MARR, A. N. (1965). 'The client's view of maintenance', Paper from RIBA Conference, *The Maintenance of Buildings 1965* (London: Ministry of Public Building and Works).
64. RIBA (1957). Conference Papers on Finance, 'Design and durability of buildings', *J. R. Inst. Bldg. Admin.* **64** (June, July, August) (London).
65. PARKER, T. W. (1965). 'Building research and maintenance', Paper from RIBA Conference, *The Maintenance of Buildings 1965* (London: Ministry of Public Buildings and Works).
66. Ministry of Works figures kindly supplied by the Economics Division, Building Research Station, Garston, England (1965).
67. CENTRAL STATISTICAL OFFICE (1968). *National Income and Expenditure* (London: HMSO).
68. *Collins English Gem Dictionary* (1961) (London: Collins).
69. *Pocket Oxford Dictionary* (4th edn. revised, 1967) (Oxford: Clarendon Press).
70. BRITISH STANDARDS INSTITUTION (1950). *Durability* CP3 (London: BSI).

71. *Report of the Committee on Building Legislation in Scotland* (1957), p. 42 (Edinburgh: HMSO).
72. GREATHOUSE, G. A., and WESSEL, C. J. (1954). *Deterioration of materials* (New York: Reinhold).
73. SCHAFFER, R. J. (1951). 'Some aspects of durability and weather', *Building Congress* (Div. 2) (London).
74. Annual Reports of the Warren Spring Laboratory Steering Committee and the Director of the Warren Spring Laboratory (London: HMSO).
75. *Climate and design* (1966). Symposium, Welsh School of Architecture (Cardiff).
76. DUCKWORTH, F. S., and SANDBERG, J. S. 'The effect of cities on horizontal and vertical temperature gradients', *Bull. Am. Met. Soc.* **35** (5), 198–207.
77. SPENCE, M. T. (1936). 'Temperature changes over short distances in the Edinburgh district', *Q. J. R. Met. Soc.*, **62**, 25–31.
78. BALCHIN, W. G. V., and PYE, N. (1947). 'A microclimatological investigation of Bath', ibid., **73**, 297–323.
79. SUNDBORG, A. (1950). 'Local climatological studies of temperature conditions in an urban area', *Tellus*, **2** (3), 2211–31.
80. KRATZNER, P. A. (1956). *Das Stadklima* (Brunswick: Friedr. Vieweg).
81. LANDSBERG, H. E. (1956). 'The climate of towns', *Man's role in changing the face of the earth*, 584–606 (Chicago).
82. PARRY, M. (1956), 'Local temperature variations in the Reading area', *Q. J. R. Met. Soc.*, **82**, 45–7.
83. CHANDLER, T. J. (1960). 'Wind as a factor of urban temperatures—a survey in North-east London', *Weather*, **15** (6), 204–13.
84. —— (1961). 'The changing form of urban temperature distribution', *Geographia*, **46**, 295.
85. —— (1962). 'Temperature and humidity traverses across London', *Weather*, **17** (7), 235–42.
86. BUILDING RESEARCH STATION (1957). 'References in wind velocities and air movement in towns', *BRS Library Bibliography*, no. 160 (London: HMSO).
87. NEWBERRY, C. W., and WISE, A. F. E. (1962). 'How wind shapes buildings—Wind effects on buildings and structures', *BRS Current Papers, Design Series*, no. 9 (London: HMSO).
88. WISE, A. F. E., and SEXTON, D. E. (1964), 'Wind tunnel studies at the Building Research Station', *BRS Current Papers, Engineering Series*, no. 15 (London: HMSO).
89. GEIGER, R. (1965). *The climate near the ground* (4th edn) (Harvard and Cambridge University Press).
90. BAILEY, J. (1968). 'Building environment, Section 1, climate and topography', *Architects J.*, **148** (40), 747–63 and **148** (41), 815–30.
91. LAWSON, T. (1968). 'Building environment. Section 2, sunlight: direct and diffused', ibid., **148** (41), 879–98; (42), 949–68; (44), 1017–36; (46), 1147–68; (47), 1215–34.

92. —— (1968). 'Building environment. Section 3, air movement and natural ventilation', ibid., **148** (48), 1283-98.
93. CHANDLER, T. J. (1965). *The Climate of London* (London: Hutchinson)
94. HAWKE, E. L. (1933). 'Extreme diurnal ranges in the British Isles', *Q. J. R. Met. Soc.*, **59**, 261-5.
95. —— (1944). 'Thermal characteristics of a Hertfordshire frost hollow', ibid., **70** 23-28.
96. PARMELEE, G. V. (1954). 'Irradiation of vertical and horizontal surfaces by diffuse solar radiation from cloudless skys', *Heat. Pip. and Air Conditioning*, **26** (8), 129-37.
97. STAGG, J. M. (1961), 'Solar radiation at Kew Observatory', *Geophysical Memoirs*, no. 86.
98. BUTEUX, H. E. (1969) 'The architect and maintenance. The relationship between design and maintenance', Paper 1, *Conf. Edin. MPBW, Div. Res. Inf.* (London: HMSO).
99. CHAPLIN, M. F. (1965). 'Analysis of maintenance', Paper from RIBA Conference, *The Maintenance of Buildings* (London: RIBA).
100. BROOKS, C. E. P. (1946). 'Climate and the deterioration of buildings, *Q. J. R. Met. Soc.*, **72** (311), 87-97.
101. —— (1950). *Climate in everyday life* (Oxford: Pergamon Press).
102. REINERS, W. J. (1962). 'Studies of maintenance costs at the Building Research Station', *The Chartered Surveyor* (March).
103. ATKINSON, B. (1970). *An aetiology of deterioration in external building materials*. PhD thesis, Queen's University of Belfast.
104. TIBBETTS, D. C. (1961). 'Some field observations of paint performance on wood sidings and trim', *DBR Tech. Note*, no. 336 (Ottawa: NRCC).
105. YAMASAKI, R. S. 'Chemical kinetics of photo-oxidative degradation of dried trilinolein film', *J. Paint Tech.*, **39** (506), 134-43.
106. BUILDING RESEARCH STATION (1959). *Principles of modern building* (London: HMSO).
107. SCAETTA, H. (1935). 'Terminologie climat, bioclimat et microclimat', *La Met.*, **11**, 342-7.
108. ELM, A. C. (1947). 'Paints as moisture barriers', *Official Digest, Fed. of Paint and Varnish Prod. Clubs*, 197-228 (April).
109. MEETHAM, A. R. (1956). *Atmospheric pollution: Its origins and prevention* and —— (1964). *The origins and prevention of atmospheric pollution* (London: Pergamon Press).
110. —— (1945). 'Atmospheric pollution in Leicester', *DSIR Technical Report*, no. 1 (London: HMSO).
111. GILPIN, A. (1963). *Control of air pollution* (London: Butterworth).
112. GARNETT, ALICE, (1957). 'Geographic factors in atmospheric pollution', *Air Pollution*, ed. Thring, M.W. Report on Conference held at University of Sheffield, Sept. 1956 (London: Butterworth).
113. HOLLAND, J. S. 'Diffusion problems in hilly terrain', *Air Pollution*, ed. McCabe, US Tech. Conf. on Air Pollution (New York: McGraw-Hill).

114. GOSLINE, C. A. et al. (1956). 'Section 5. The evaluation of weather effects', *Air Pollution Handbook*, ed. Magill, P. L., et al. (New York: McGraw-Hill).
115. SMITH, M. E. (1951). 'Meteorological factors in atmospheric pollution problems', *Am. Inst. Hyg. Assoc. Q.*, **12**, 151–4.
116. HEWSON, E. W. (1945). 'The meteorological control of atmospheric pollution', *Q. J. R. Met. Soc.*, **71**, 266–82.
117. VERNON, W. H. J. (1935). 'A laboratory study of atmospheric corrosion of metals. Pts. 2–3; *Trans. Faraday Soc.*, **31**, 1668.
118. —— (1943). 'Jubilee Memorial Lecture on corrosion. Society of Chemical Industry', *Chem. and Ind.*, **62**, 314–18.
119. —— *Trans. Faraday Soc.* **23** (1927) 113; **27** (1931) 255; **31** (1935) 1668; *Trans. Electrochem. Soc.*, **64** (1933) 31.
120. —— (1949). 'Corrosion of metals (I, general principles, II preventative measures), *J. R. Soc. Arts*, **97**, 578–610.
121. EVERETT, L. H., and TARLETON, R. D. J. (1967). 'Recognition of corrosion hazards to metals in building', *J. Brit. Corros.* (2), 61–4.
122. SHELTON, H. A. (1965). 'Accelerated weathering for metal finishes', *Trans. Inst. Metal Fin.*, **43** (5), 179–85.
123. SCHIKORR, G. (1948). 'Atmospheric corrosion of metals', *Arch. Metallkunde*, **2**, 223–30.
124. IRON and STEEL INSTITUTE (1966). Conference on *Corrosion at Joints in Constructional Materials* (London).
125. HUDSON, J. C. (1948). 'Atmospheric corrosion of ferrous metals and its prevention', *J. Iron and Steel Inst.* **160**, 276–86.
126. —— (1940). *The corrosion of iron and steel* (London: Chapman).
127. —— and STAMNERS, J. F. (1953). 'The effect of climate and atmospheric pollution on corrosion', *J. Appl. Chem.*, **3** (2), 86–96.
128. —— —— (1955). 'The corrosion resistance of low-alloy steels', *J. Iron and Steel Inst.*, **180**, 271–89.
129. SEREDA, P. J. (1960). 'Measurement of surface moisture and sulphur dioxide activity at corrosion sites', *ASTM Bulletin*, no 246, pp. 47–8.
130. —— (1961), 'Characteristics of moisture deposition on corrosion specimens', *Research Paper*, no. 146 (Ottawa: NRCC).
131. OMA, K., et al (1965). 'Studies on atmospheric corrosion of steels related to meteorological factors in Japan', *Corrosion Engineering (Boshoku Gijyutsu)*, **14** (1), 16.

Chapter 2 CONCRETE

J. GILCHRIST WILSON

FRIBA
*Concrete Finishes Consultant and formerly
Senior Advisory Architect to the
Cement and Concrete Association*

INTRODUCTION

Concrete is a generic name for a conglomerate consisting of fine and coarse aggregate, bonded together by a paste composed of Portland cement and water.

By altering the type of aggregate used in the mix it is possible to produce a range of concretes suitable for a wide variety of purposes; this range can be enlarged still further by the use of cements other than Portland and by the use of reinforced and prestressed concrete. It is unique among major structural materials in that it can be precast in a factory, or made on site; it imposes no restrictions on structural form other than those enforced by the limitations of practicability or the skill of the designer. It is a material about which a vast amount of literature has been produced embracing a number of standard works of reference together with an infinite number of reports and papers on every conceivable aspect of the subject.

The aim of this chapter has been to bring together in as concise a manner as possible the basic information about the factors which influence the performance and weathering of concrete.

FACTORS AFFECTING PERFORMANCE

General

The performance and durability of concrete depend very largely upon its resistance to deterioration and the environment in which it is placed, which in turn depend upon the quality of the constituent materials of which the mix is composed, and the manner in which these materials are proportioned, mixed, placed, compacted, and cured.

Generally the cement used should comply with one or other of the following British Standards.[1] BS 12, BS 146, BS 4027, or BS 915 and the aggregates with BS 882 or BS 1047 except for lightweight aggregates which should comply with either BS 3797 or BS 877. The mix proportions should be selected to ensure that the workability of the fresh concrete is suitable for the conditions of handling and placing, so that after compaction the concrete completely surrounds all reinforcement and entirely fills the formwork.

Apart from durability the performance of concrete is affected by another component, namely 'deformation' which relates to the natural laws governing the behaviour of materials.

Durability

The resistance of concrete to frost and chemical action depends very largely upon its quality and those of the constituent materials. In reinforced and prestressed concrete, corrosion of the steel reinforcement is influenced by the depth of cover provided and the permeability of

[1]. For list of British Standards referred to in this chapter see Appendix 1.

Concrete 43

the concrete. In the great majority of environments, adequate durability can be achieved by using a concrete having a cement content of not less than 330 kg/m³ and by using a water/cement ratio which will provide a mix that is just sufficiently workable to enable it to be fully compacted.

Aggregates having a high drying shrinkage, such as some dolerites and gravels, produce concrete having a higher drying shrinkage than normal. This results in deterioration of exposed concrete and excessive deflections of reinforced concrete unless special measures are taken(1).

Permeability   The permeability of concrete influences its resistance to frost action and chemical attack, and more important still, the protection of reinforcing steel against corrosion.

For given aggregates the permeability of concrete can be reduced by increasing the cement content or by lowering the water content; there is, however, a limit in practice to the reduction of water content which can be made if full compaction is to be achieved.

Table 2.1 based on the draft BS 'Code of Practice for the structural use of concrete' (2), and CP 116, gives the minimum cement content of fully compacted concrete and the nominal cover for various grades of concrete to meet particular conditions of exposure.

Cover to reinforcement   The nominal concrete cover given in Table 2.1 applies to all reinforcement and tendons, including stirrups and links, and is that to be shown on working drawings. The nominal cover associated with different grades of concrete applies only to concretes made with aggregates complying with BS 882 or BS 1047 (see Appendix).

Frost resistance   The number of cases in England and Wales where concrete used as the exposed vertical face of a structure has been

Table 2.1   *Grades of fully compacted concrete for strength and durability purposes using cement complying with BS 12 or BS 146.*

PART A. Selection of the grade of concrete necessary to meet strength requirements

Grade	Characteristic cube strength N/mm^2 for the following grades						
	22·5	30	37·5	45	52·5	60	67·5
Age at test							
28 days	22·5	30·0	37·5	45·0	52·5	60·0	67·5
7 days	15·0	20·0	26·0	32·0	38·0	45·0	52·0

PART B. Selection of the grade of concrete, minimum cement content and nominal cover necessary to meet durability requirements.†

	Minimum cement content of fully compacted concrete		Nominal cover for grade			
	Reinforced kg/m^3	Prestressed kg/m^3	22·5 mm	30 mm	37·5 mm	45 and over mm
Conditions of external exposure‡						
Sheltered against direct rain and against freezing whilst saturated with water	280	300	40	30	25	25
Exposed to alternate wetting and drying and to freezing whilst wet	330	330	NA*	40	30	25
Exposed to sea water	370	370	NA	NA	60	50

*NA indicates that this grade of concrete is not allowed under the particular conditions of exposure.
†Attention is drawn to *BRS Digest*, no. 109 (1969), 'Zinc-coated reinforcement for concrete'.
‡For a quantitative interpretation of these terms attention is drawn to *BRS Digest*, no. 23 (1962), 'An index of exposure to driving rain'.

damaged by frost action is relatively rare. Only concrete possessing a particular size of pore structure usually brought about by the use of too much water in the mix, is likely to be affected, and then only if the concrete happens to be in a high degree of saturation when the frost attack takes place. Air-entrained concrete has a high resistance to frost attack due to the formation of a large

Concrete

number of air bubbles within the concrete. The amount of entrained air in the matrix of the concrete should be of the order of 13 per cent, which means that for a 10 mm maximum size coarse aggregate a typical air content would be 7 per cent (of concrete volume) and for a 38 mm size aggregate, 4 per cent.

The inclusion of air in the mix naturally reduces the strength of the concrete, but as it makes for greater workability this means that, in practice for similar strength and workability, air-entrained concrete will require a cement content of approximately 4 per cent more than for plain concrete.

Resistance to chemical attack

The most common form of chemical attack is that due to sulphates of calcium, magnesium, sodium, and potassium which occur widely in clays and in ground-waters, and as such can be ignored for the purpose of this study. Ample data is available covering the requirements for concrete exposed to sulphate attack (3).

Structural considerations

As indicated in the Introduction, concrete is unique amongst building materials in being able to satisfy the most visionary concepts of the designer's imagination—viz. the Sydney Opera House (Fig. 2.1)—whilst at the

Fig.2.1. Sydney Opera House, New South Wales, Australia. An imaginative but expensive use of precast post-tensioned concrete. *Architect:* Joern Utzon. *Engineers:* Ove Arup & Partners. *Photograph:* J. R. F. Stewart.

same time fulfilling all the normal functions of a structural material. This freedom has, quite naturally, had a profound influence on the architecture of this and other countries; unfortunately not all designers have appreciated that like other more traditional building materials concrete has certain practical limitations which if not taken into account at the design stage can, and frequently do, have an influence on the subsequent performance and weathering behaviour of the concrete. The most obvious limitation is the inability of even the most co-operative of contractors to cast all the concrete in one pour; this means that construction joints have to be provided. It is the duty of the designer to work out with the other members of the building team the location and design of all such joints. This may sound an obvious comment to make but it is surprising how often it is overlooked and the appearance of the structure is marred by the unhappy disposition of the joint pattern. Another important limitation is that, like other materials, concrete is subject to movements as a result of deformation. Where these movements are sufficiently large they can result in visible cracking and/or foreshortening of the structural members so that damage is done to finishes, claddings, partitions, etc.

Deformation

The principal components of the deformation of concrete are:
1. Elastic deformation which takes place instantaneously;
2. Shrinkage which occurs over a long period and is independent of the stress in the concrete;
3. Creep which also occurs over a long period but is dependent on the stress in the concrete.

Unfortunately the numerical values which need to be taken into account when considering the above three components in design, depend upon a large number of factors including the effects of environment, age of the

Concrete

concrete, its mix proportions, and its constituent materials.

Pending the publication of the BS 'Code of Practice for the structural use of concrete' (2), reference should be made to CP 115, *The structural use of prestressed concrete in buildings*, which gives values of modulus of elasticity of Portland cement concrete where aggregates comply with BS 882. Typical average values for modulus and elastic strain are shown in Table 2.2.

Table 2.2

Cube strength at 28 days N/mm^2	Modulus of elasticity E (kN/mm^2)	Elastic strain per unit stress per mm length $\times 10^{-6}$ Reinforcement (%)			
		0·1%	0·4%	1·0%	4·0%
22·5	22	45	44	42	35
30·0	25	40	39	37	30
37·5	28	36	35	33	27
45·0	30	33	32	31	25

Shrinkage

The shrinkage of concrete is dependent on the amount of drying that can take place. In thin unreinforced sections strains may be of the order of 500×10^{-6} which represents an unrestrained shrinkage of the order of 1·5 mm per 3 m length of a concrete member. Shrinkage is closely dependent on the cement paste and the moisture content of the concrete mix.

Under normal climatic conditions the rate of shrinkage may be assumed to be:

After 14 days	30 per cent of total shrinkage
After 28 days	40 per cent ,, ,,
After 3 months	60 per cent ,, ,,
After 1 year	80 per cent ,, ,,

Creep

Creep of concrete is dependent on the stress in the concrete. For stresses of up to one-third of the cube strength, it may be assumed that it is directly proportional to stress.

Creep of concrete under stress tends to reduce the maximum stresses arising from thermal and drying shrinkage; it is a long-term process and if the stresses change rapidly due to a change in the cross-sectional area of the member permitting its temperature and shrinkage to take place in a relatively short time, it will have little value in reducing stresses. Creep in itself is a factor which has been responsible for a great deal of cracking of partitions and to numerous failures of external cladding due to the foreshortening of columns, etc., under loading. It may be assumed that half the total creep takes place in the first six months after loading.

Coefficient of thermal expansion

The coefficient of thermal expansion varies according to the type of aggregate used in the concrete mix, although the paste content, age of concrete, and relative humidity will also have a slight influence upon the value. Typical values for concrete using different types of aggregate over the normal temperature range are given in Table 2.3.

Table 2.3 *Coefficients of thermal expansion of various concretes*

Aggregate type	Coefficient of thermal expansion per °C
Gravel	$12 \cdot 6 \times 10^{-6}$
Granite	$9 \cdot 4 \times 10^{-6}$
Sandstone	$11 \cdot 7 \times 10^{-6}$
Limestone	$7 \cdot 2 \times 10^{-6}$
Dolerite	$9 \cdot 4 \times 10^{-6}$

From the foregoing it will be seen that there are numerous factors influencing the tendency for concrete to shrink and crack; the limitation of such cracking is also influenced by many factors, probably the most important of which is the proper provision of adequate reinforcement. To distribute cracking arising from shrinkage and thermal stresses in structures which are normally unreinforced,

Fig.2.2. The college shown in this picture illustrates the weathering pattern of the precast concrete after some seven years of exposure in a comparatively clean atmosphere.
Architects: Sir Basil Spence, Bonnington & Collins. *Engineers:* Ove Arup & Partners. *Precast concrete:* Modular Concrete Co Ltd. *Photograph:* J. G. Wilson.

particularly those directly exposed to the weather, the total reinforcement should be not less than 0·4 per cent of the volume of the concrete in the wall. The reinforcement should be placed near the exposed surfaces and with the specified cover given in Table 2.1. On vertical faces it should be dispersed two-thirds horizontally and one-third vertically.

FACTORS AFFECTING APPEARANCE

General

Development work carried out over the past two decades into ways and means for texturing and otherwise treating the surface of concrete, both *in situ* and precast, has largely been responsible for today's improvement in the image of concrete. Hand in hand with these developments has been a steady increase in the perceptive use of exposed 'board-marked' concrete; unfortunately many early examples of this concept displayed blemishes which at the time were unaccountable; however, research has now established the cause of most of these blemishes as well as the means for overcoming them.

The Table 2.4, taken from the Cement and Concrete Association's publication *Recommendations for the production of high quality concrete surfaces* (4) defines most of the common blemishes that occur on concrete surfaces and lists their most probable causes. It is hoped that the building industry will accept the terminology and definitions used in this classification as a means of identification, and to facilitate communication.

Table 2.4   *Definitions of blemishes on concrete surfaces and their most probable causes.*

A. COLOUR VARIATIONS

1. Visible immediately or within a few hours of striking the formwork

Description	Common term	Most probable causes	
Variation in colour of the surface	Inherent colour variation	Materials	inconsistent grading or colour
		Concrete mix	incomplete mixing, segregation or variation in proportions
Dark areas of size and shape similar to the coarse aggregate. Mottled appearance	Aggregate transparency	Formwork	too flexible
		Concrete mix	low sand content, gap graded
		Placing methods	excessive vibration, external vibration
Light areas of size and shape similar to the coarse aggregate. Mottled appearance	Negative aggregate transparency	Materials	aggregate dry or highly porous
		Curing	too-rapid drying
Variation in shade of the surface	Hydration discoloration (due to moisture movement within or from the fresh concrete)	Formwork	variable absorbency: leakage through joints
		Release agent	uneven or inadequate application
		Curing	uneven

Concrete

Description	Common term	Most probable causes	
Variation in colour or shade, giving a flecked appearance	Segregation discoloration (separation of fine particles due to bleeding parallel with the form face)	Formwork Concrete mix Placing methods	low absorption lean excessive vibration: low temperature
Discoloration foreign to the constituents of the mix	Dye discoloration	Formwork Release agent Materials	stains, dyes or dirt on the form face impure dirty
Cream or brown discoloration. Sometimes showing sand or aggregate	Oil discoloration	Release agent	excessive, low viscosity, impure, applied too late
Light-coloured dusty surface which may weather to expose aggregate	Dusting	Curing Cement Release agent	inadequate (very rapid drying) air-set excessive application of chemical agent
Matrix near the colour of sand and lacking in durability	Retardation	Formwork Release agent	retarder in or on form face water soluble emulsion: cream or oil with excessive surfactant: unstable cream: unsuitable or excessive chemical release agent

2. Visible only some time after striking the formwork

Description	Common term	Most probable causes	
Variation in shade of the surface, from light to dark	Drying discoloration	Curing	different conditions
White powder or bloom on the surface	Lime bloom or efflorescence	Design Release agent Curing	permitting uneven washing by rain type uneven conditions
Discoloration foreign to the colour of constituent materials	Contamination	Materials Reinforcement Curing	pyrites, clay or other impurities inadequate cover: rust from steel above impure curing compounds: dirty covers

B. PHYSICAL IRREGULARITIES

1. Visible immediately after striking the formwork

Description	Common term	Most probable causes	
Coarse stony surface with air voids and lacking in fines	Honeycombing	Concrete mix Formwork Placing methods Design	insufficient fines, too low workability joints leaking causing segregation: compaction inadequate highly congested reinforcement: too narrow section
Individual cavities usually less than $\frac{1}{2}$ in dia. Small cavities approximately semi-spherical: larger cavities often bounded by stone particles	Blowholes	Formwork Release agent Concrete mix Placing methods	form face impermeable, with poor wetting characteristics: inclined: too flexible neat oil without surfactant too lean, too coarse a sand, too low a workability inadequate compaction: too slow rate of placing: external vibration
Sand textured areas, devoid of cement. Usually associated with dark colour on adjoining surface	Grout loss	Formwork	leaking at joints, tie holes, etc.
Irregular eroded areas and channels having exposed stone or sand particles	Scouring	Concrete mix Placing methods	excessively wet: insufficient fine particles: too lean water in formwork: excessive vibration on wet mix: low temperature when placing
Step, wave or other deviation from the intended shape	Alinement or profile variation	Formwork Placing methods	damaged: deformed under load: joints not tightly butted too rapid or careless
Short cracks often varying in width along their length. With vertical casting, cracks more often horizontal than vertical	Plastic cracking	Concrete mix Formwork	high water/cement ratio: low sand content poor thermal insulation: irregular shape (restraining settlement)

Concrete

2. Visible immediately or some months after striking the formwork

Description	Common term	Most probable causes	
Thin layer of hardened mortar removed from the concrete surface, exposing mortar or stone	Scaling	Formwork	relaxing after compaction: form face excessively rough
		Release agent	inadequate application or removed during subsequent operations: ineffective
		Concrete mix	low strength
		Striking time	too early
Parts of the form face, including barrier paint, adhering to the concrete	Form scabbing	Formwork	form face excessively rough, weak or damaged
		Release agent	inadequate application or removed during subsequent operations: ineffective
		Striking time	too late
Pieces of concrete removed from the hardened surface. Deeper and usually more severe than scaling	Spalling or chipping	Formwork	difficult to strike
		Release agent	inadequate application or removed during subsequent operations: ineffective
		Concrete mix	low strength: aggregates susceptible to damage by frost or water
		Striking time	too early: mechanical damage after striking
		Weathering	frost action: corrosion of reinforcement
A network of fine cracks in random directions, breaking the surface into areas from about ¼—3 in across	Crazing	Formwork	form face of low absorbency, smooth or polished
		Concrete mix	too rich in cement, too high water/cement ratio
		Curing	inadequate
		Striking time	too early, especially in cold weather

Variations in colour   The colour of concrete depends upon the colour of the cement and fine aggregate used in the mix. The influence of the finer particles of sand becomes more pronounced with the passage of time, as more and more of the cement is lost from the face of the concrete by the action of the elements. In aggressive atmospheres it is not uncommon for the surface of concrete to take on the appearance of an acid-etched finish in a matter of a few years, hence, in those instances where colour is of extreme importance, the need for using a sand similar in colour to that of the cement.

In the south of England most of the natural sands are yellowish in colour, which accounts for the reason why so much of the concrete in that part of the country takes on an unattractive yellowish tone after a few years of weathering.

The importance of ensuring that the source of supply of both the cement and that of the aggregates remains constant throughout the contract cannot be too strongly emphasized (see Appendix 2).

Other reasons for variations in the colour of concrete are brought about by inconsistancies in the mix proportions—changes in the water content from batch to batch can influence the final colour of the concrete quite appreciably.

The time interval between the placing of the formwork and the stripping of the concrete can also have an effect on the colour of the concrete; every endeavour should be made to keep the time of stripping formwork constant throughout that part of the work where colour is important.

Possibly the factor reponsible for the greatest colour contrasts in cast *in situ* concrete result from what is now known as hydration discoloration. It was not until the Cement and Concrete Association had carried out its investigation into the causes and reasons for the common blemishes on concrete surfaces, that it was established that loss of moisture through joints between boards or formwork panels caused a distinct change in the colour of the concrete, and possibly more important still, it was shown that the colour of the concrete could be related to the absorbancy of the formwork.

Concrete

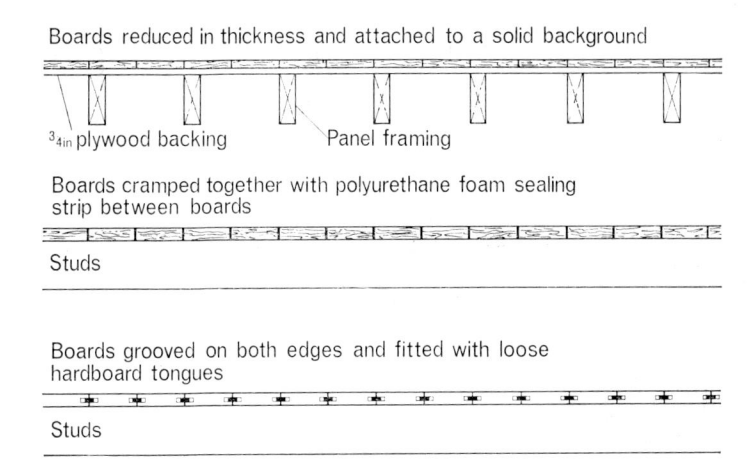

Boards reduced in thickness and attached to a solid background

$^3{}_{4in}$ plywood backing   Panel framing

Boards cramped together with polyurethane foam sealing strip between boards

Studs

Boards grooved on both edges and fitted with loose hardboard tongues

Studs

Fig.2.3. Three methods of preventing leakage of moisture at board joints in formwork.

Fig. 2.4. Methods of forming construction joints in formwork.

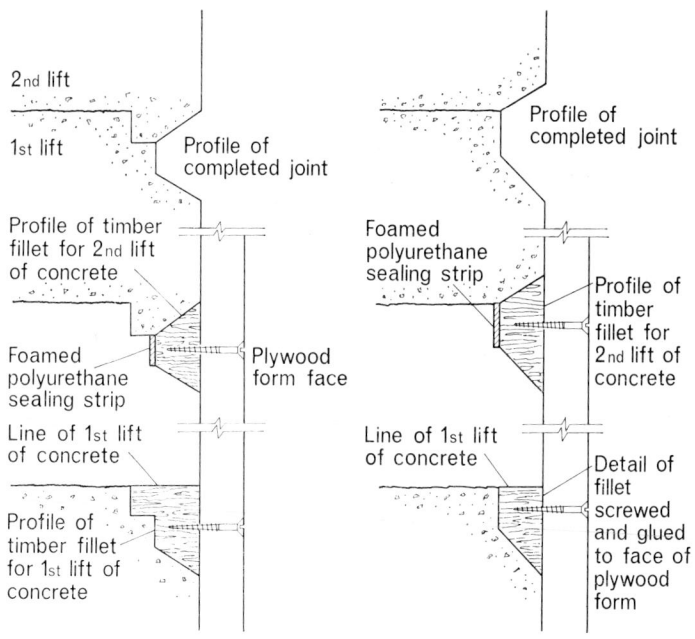

2nd lift
1st lift
Profile of completed joint
Profile of timber fillet for 2nd lift of concrete
Foamed polyurethane sealing strip
Plywood form face
Line of 1st lift of concrete
Profile of timber fillet for 1st lift of concrete

Profile of completed joint
Foamed polyurethane sealing strip
Profile of timber fillet for 2nd lift of concrete
Line of 1st lift of concrete
Detail of fillet screwed and glued to face of plywood form

Fig.2.5 (a). Queen Elizabeth Hall, Purcell Room and Hayward Gallery, South Bank, London, SE 1. Constructed entirely of concrete this complex illustrates a variety of finishes. The weathering of some of the surfaces shows the need for a careful study of design detailing. *Architect:* Hubert Bennett, architect to the Greater London Council. *Engineers:* Ove Arup & Partners. *Contractors:* Higgs & Hill Ltd. *Photograph:* J. G. Wilson.

Fig.2.5 (b). Detail of rough board finish used on Queen Elizabeth Hall illustrated in Fig. 2.5 (a).

Fig.2.5 (c). The method used on the Queen Elizabeth Hall to prevent nail-heads or other fixings from showing on the face of the concrete.

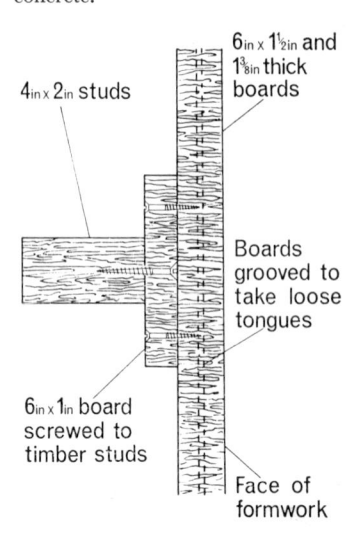

4 in x 2 in studs

6 in x 1 in board screwed to timber studs

6 in x 1½ in and 1⅜ in thick boards

Boards grooved to take loose tongues

Face of formwork

Concrete

Hydration discoloration

Discolorations of this type are caused by moisture movements within or from the fresh concrete. The most common form of hydration discoloration occurs when moisture is permitted to escape at board or panel joints. Another cause of discoloration is the result of moisture being absorbed into the body of the formwork; when this takes place, the water which bleeds from the concrete face brings with it the finer cement particles, resulting in a surface having a high cement content and low water content, which in turn means a high concentration of tetracalcium aluminoferrite (C_4 Af), the chemical which gives grey cement its colour. The final result is a darkening of the colour of the concrete over the area from which moisture has been lost—this darkening can vary in shade from a fairly darkish grey to almost black.

This change in colour due to loss of moisture from the fresh concrete is clearly demonstrated where concrete has been cast against unsealed plywood panels, the veneers of which vary between heartwood and sapwood, and in some cases the sheets are composed of mixed strips of veneers, consequently considerable variations in absorbancy occur across each ply sheet and from sheet to sheet. These variations produce corresponding variations in the colour shade of the concrete.

Yet another form of hydration discoloration is due to movements of the form face prior to stripping—if the

Fig.2.6. Hydration discoloration caused by loss of moisture from the concrete mix through joints between formwork boards. Joints between boards and/or panels must be watertight.
Photograph: S. W. Newbery.

form face loses contact with the face of the concrete, as frequently happens when a form liner is used, this will permit differential moisture movements from the face of the concrete resulting in a change in the colour of the concrete.

Segregation discoloration

Murphy (5), describes this phenomenon as follows:

'Segregation discoloration is the term given to the differences in colour over a surface due to the mix proportions of the concrete varying under the influence of segregation caused by the action of compacting; it does not include segregation due to other factors such as the absorption of the form face. The two basic forms of segregation discoloration are due to movements of the mix constituents parallel and perpendicular to the face of the form.

Segregation parallel to the face of the form is manifested by a flecked appearance resembling two colours that are incompletely mixed and of varying shade. The colours are those of the cement and the fine aggregate in the mix while the depth of colour is a function of the water content.

Segregation perpendicular to the face of the form causes a blemish which is termed aggregate transparency; the surface of the concrete is marked with dark areas similar in size and shape to the particles of coarse aggregate in the mix. There have been several explanations for this blemish but it is almost certainly caused by segregation during compaction of the concrete. If concrete is vibrated the finer materials tend to be drawn towards the source of vibration, that is, towards the point at which the amplitude of vibration is greatest.

If an immersion vibrator is used to compact concrete, the form face tends to vibrate with a higher amplitude than the concrete a short distance from it and fine material thus tends to be drawn to the face of the form. However, segregation must have two components because fine material can only migrate in one direction if the coarse particles move in the other direction. Thus, if concrete is vibrated, particles of coarse aggregate near the form face will tend to move slightly from it and fine material will simultaneously fill the area between each particle and the face of the form. The movement is, however, very small so that only the finest of the mortar

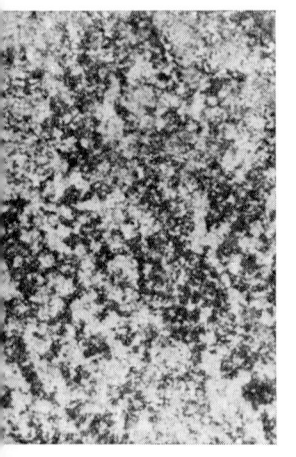

Fig.2.7. Segregation discoloration caused by variation in water content over the surface due to segregation within the mix.
Photograph:
S. W. Newbery.

Concrete

constituents can flow in this way. This means that, at the completion of vibration, the area between the coarse aggregate particle and the face of the form will be filled with a fine mortar which is rich in cement and so will appear as a dark discolouration similar in size and shape to the coarse aggregate.

Although it is suggested that there are only two basic types of segregation discolouration, neither will normally occur without the other and, in practice, both often occur to a significant degree over the same area. Segregation discolouration is also often masked by other blemishes particularly if absorbent form faces, which cause a migration of cement-rich material to the surface, are used.'

Crazing　　This is an irregular network or map pattern of superficial cracks of minute width. Although of no structural significance the crazing of concrete surfaces is a blemish having some aesthetic significance. As a general rule crazing is associated with smooth surfaces and mixes rich in cement and fine material. With mixes of this type the cement and fine material migrate under compaction to the face and form what is known as a laitance skin on the surface of the concrete. This skin, due to its composition, is very susceptible to moisture movements and under conditions of wetting and drying in association with low humidity and low temperature it breaks down into an irregular network of fine cracks. The conditions necessary to produce crazing may take place shortly after casting whilst the matrix is weak and unable to resist the stresses, or it may not occur until the concrete is twelve or more months old. No hard and fast rule can be given as to when crazing will take place but normally it takes place within the first few months after casting.

The ideal conditions for producing a craze-prone surface are to cast the concrete against non-absorbent formwork using a mix containing a high percentage of very fine material and an excess of water.

It is often stated that white cement is more prone to craze than ordinary grey Portland cement and research carried out for the Cast Stone and Cast Concrete Products Industry (6) indicates that white cements do, in fact, exhibit larger wetting and drying movements. There is, of course, the other factor that dirt-filled craze cracks are far more noticeable against a light background, nevertheless there does appear to be unquestionable proof that white cement due to the reasons given above and possibly to the fact that they contain little or no tetracalcium aluminoferrite are more craze-prone than grey cements.

The only really positive way to overcome crazing is either to prevent the rich skin of laitance from forming on the surface of the concrete, or to treat the surface to remove the skin of laitence after it has formed.

Release agents   There are a variety of release agents (mould oils) on the market, but where appearance is important the type used should be either a chemical release agent, a cream emultion, or a neat oil with surfactant (wetting agent) added. Until quite recently it was usual to blame virtually all the blemishes that occurred on the face of concrete upon the 'mould oil'. It has been shown that staining due to oil itself is relatively rare (7), in fact, it only occurs when an excessive amount of oil has been used, or when the oil has become contaminated. The study of release agents has brought to light the fact that some oils cause blowholes, others affect the density and durability of the surface and indirectly cause colour variations due to differential wetting and drying of the surface, while others retard the set of the cement in contact with them, producing a mild form of exposed aggregate finish.

The action by which the type of oil affects a concrete surface depends largely upon whether or not a surfactant has been added to the oil and upon the properties of the surfactant. The presence of a surfactant, at the interface between the concrete and the form reduces the surface tension so that the air bubbles are provided with a larger contact area with the form and can move more freely upwards and disperse.

An absorbent lining may sometimes absorb air bubbles but in any case it provides a greater contact area than does an impermeable lining and so allows the bubbles to disperse more easily.

In order to obtain a satisfactory release of the formwork, only a thin film of oil or emulsion is required; generally the coverage should be in the region of 300–400 ft^2/gal (6–8 m^2/l). The spread of a chemical release agent may be considerably greater—up to 1000 ft^2/gal (20 m^2/l)—depending upon the type of form lining, and the number of reuses it has been given. If a painted mould or form face is used, care must be taken to ensure that the release agent does not cause the paint to soften or deteriorate in any way.

Lime bloom The two main chemicals in Portland cement—grey or white—are tricalcium silicate and dicalcium silicate. When these hydrate they result in the liberation of free lime—calcium hydroxide—which can further react with carbon dioxide in the atmosphere or, dissolved in the water, to form an insoluble deposit of calcium carbonate—chalk—and it is this deposit which should be referred to as lime bloom and not efflorescence as so frequently happens; it may be removed by washing the affected areas with a dilute hydrochloric acid solution—5 per cent is usually found to be strong enough, followed by copious washing with clean water. Lime bloom is most troublesome when it appears on surfaces which in themselves are darker than the calcium carbonate. On white cement concrete it is rarely objectionable. Although lime bloom is classified as a blemish, the carbonation of free lime can be advantageous in that it is responsible for reducing the porosity of concrete by the formation of insoluble calcium carbonate within the capillaries, voids, and pores of the matrix in concrete. By the same process carbonation can in certain circumstances heal autogeneously fine craze cracks in a concrete surface.

Efflorescence This is an unsightly white stain caused by the migration in solution of soluble salts, principally calcium sulphate, to the surface where they crystallize; it is a blemish which affects other materials more than concrete. When it appears on concrete it is usually due to contact with soils or materials containing sulphates. It is a blemish which is normally dissolved by rainwater; on the other hand it can often be removed by brushing the affected areas.

Algae and other growths There are a great many varieties of algae, but those which occur on moist external concrete surfaces are usually green, red, or brown powders or filaments which may or may not be slimy according to moisture conditions. They derive their energy largely from sunlight but need liquid water for survival.

Fungi (mildew, moulds, yeasts). Unlike algae, sunlight is not an essential requirement for the growth of fungi; they depend upon organic material such as decaying plant life for their energy. Moulds normally appear as spots or patches which may spread to form a grey-green, black, or brown furry layer on the surface.

Lichens are a combination of certain algae and fungi which can readily reproduce on moist external concrete surfaces. They have great resistance to extremes of temperatures and drought; they are slow in growth but long lived. Lichens are sensitive to oxides of nitrogen and sulphur, hence the absence of lichens in industrially polluted atmospheres.

Mosses need a rough moist surface on which soil and dirt can collect; once established they tend to hold moisture in the supporting surface.

Concrete

Toxic washes can be used to kill algae, lichens and mosses on external concrete surfaces (8) and below is a list of firms manufacturing suitable materials for this purpose, together with the trade name of their particular product.

Lithurin	Chemical Building Products Ltd.
Topane WS	ICI Ltd, PO Box 19, Templar House, High Holborn, London, WC1.
Shirlan NA	ICI Ltd, PO Box 19, Templar House, High Holborn, London, WC1.
Panacide	BDH, Poole, Dorset.
Nubex SF	Nubold Development Ltd, Crawley, Sussex.
Santobrite	Monsanto Chemicals Ltd, Victoria Street, London, SW1.

White cement concrete

Many architects consider that the colour of concrete *is grey* and no attempt should be made to produce concrete of any other colour: there are, however, others who dislike the particular cold grey colour of concrete and consider it quite legitimate to substitute white cement for ordinary Portland cement in the mix, contending that by so doing they obtain far greater value from the light-reflecting properties of the surface and at the same time overcoming one of the very real drawbacks of normal grey cement concrete, namely the violent contrast in colour that results between wet and dry surfaces following rain.

White cement complies in every respect with BS 12—this is not always fully appreciated, or altogether believed, nevertheless it is a fact.

The choice of aggregates to use with white cement depends very largely upon the type of finish it is proposed to give to the concrete. Under 'Variations in colour' it was pointed out that the resulting colour of any concrete depends mainly on the colour of the cement and to a lesser degree on the colour of the fine aggregate used in the mix. If the concrete is to be left direct from the formwork without any subsequent treatment of the surface, the colour of the coarse aggregate is relatively unimportant, except possibly in a highly aggressive atmosphere. If, on the other hand, the surface of the concrete is to be tooled, or treated in any other way to remove the skin of cement mortar from the face, then the choice of both the fine and the coarse aggregate becomes equally important.

The range of aggregates in the British Isles suitable for use with white cement to produce a white or light-coloured concrete is restricted to the lighter-coloured limestones and granites, calcite spar, calcined flint, and to one or two desposits of natural fine silica sand. The choice for any particular purpose will depend primarily on the cost of the materials, the availability of supplies, and in certain circumstances on the particular quality of 'whiteness' required.

The mix requirements for white cement concrete are fundamentally the same as those for ordinary Portland cement concrete, the main difference being that in the case of ordinary Portland cement concrete it is customary to specify that both the fine and the coarse aggregate shall comply with the grading limits laid down in BS 882. In the case of white cement concrete, it is frequently not economically possible to obtain a light-coloured fine aggregate that will comply with any of the grading zones laid down in BS 882, the main reason for their non-compliance being the percentage of fine material that will pass the No. 52 sieve. However, this does not mean that a perfectly satisfactory concrete cannot be produced, but what it does mean is that the percentage of sand required to give the necessary workability and cohesiveness to the mix may have to be reduced below that usually used. In other words the mix will need to be designed in accordance with Clause 209 rather than Clause 208 of CP 114.

The surface finish required will affect the choice of mix depending upon whether the aggregate is to be exposed or the concrete left untreated from the formwork. If the aggregate is to be exposed, the degree and type of exposure is important as this will largely determine whether a continuously graded or a gap-graded mix is the most suitable. Again, the type of exposure—e.g. bush-hammering or abrasive blasting—normally calls for the use of different mix proportions (see Appendix 2).

Where the finish required is a white concrete left untreated from the formwork, the designer is mainly concerned with the grading and colour of the fine aggregate, the choice of which will to a large extent be governed by cost and availability of supplies. The coarse aggregate, i.e. material above $\frac{3}{16}$ in (5 mm), can within reason be any hard, durable, clean, natural gravel or crushed stone. In the case of a bush-hammered or a point-tooled finish, both fine and coarse aggregate must be light in colour and it will generally be found that a continuously graded aggregate is the most suitable for finishes of this type.

For deeply exposed aggregate finishes such as those obtainable by early stripping of the formwork and then washing and brushing the surface, or aggregate-revealing by abrasive blasting, the best results are obtained by using a gap-graded concrete mix in which the coarse aggregate is all retained on a $\frac{3}{4}$-in (20 mm) sieve. The fine aggregate will normally be either $\frac{3}{16}$ in (5 mm) down for natural pit sand, or $\frac{1}{8}$ in (3 mm) down for a crushed rock fines. For the best results, the grading should wherever possible come within the limits of Zone 3 or 4 of BS 882.

To establish the mix proportions for a required finish it is clearly desirable to make a number of trial mixes; then, having established the most suitable mix, to cast a sample section of a wall and to treat the surface as required. Having once established the mix proportions it is

Fig.2.8. The American Lutheran Church, Oslo. White cement and white spar aggregate were used in this example of 'Naturbetong'.
Photograph: J. G. Wilson.

essential to make sure that the quarry or pit can supply adequate quantities of materials to a constant grading.

Much of the above material comes from the booklet *White concrete with some notes on black concrete* (9) published by the Cement and Concrete Association.

FACTORS AFFECTING WEATHERING

General

In any study dealing with the performance and weathering of concrete it must be appreciated that whilst natural materials like stone have required millions of years to achieve their homogeneous nature and make them fit for building purposes, concrete can harden in a matter of hours, become self-supporting in a matter of days, and develop its full potential in a matter of weeks. Thus, partly because of the short time it is given to mature, but largely because of its heterogeneous composition, concrete will always be sensitive to the effects of weathering, particularly in the early stages of its life. In fact quite a number of designers who have studied the behaviour of concrete have come to the conclusion that it takes concrete about ten years to reach adolescence and that any pronouncement on the quality of weathering made before the structure has reached this age is to prejudge the issue.

The passage of time leaves its imprint on every building, irrespective of its architectural merits. Unfortunately, unless designed to do so, not every building bears these traces happily, or even improves in appearance under their influence. The contrast between the discoloured and rain-washed parts of a building, resulting in an emphasis of its main features and a reinforcement of its architectural style, can be seen in some buildings: whether this was a conscious effort on the architect's part or merely fortuitous it is difficult to say: nevertheless it is obvious that it

Fig.2.9. The concrete water tower with sun room above reached by an outside staircase, showing the effects of forty years of weathering.
Architect: Maxwell Ayrton. *Photograph:* J. G. Wilson.

Fig.2.10. Twickenham Bridge, Surrey. This illustration shows the effect of some thirty-three years of weathering. The vertical streaking caused by the run-down of moisture has destroyed the unity of the bush-hammered concrete surface.
Architect: Maxwell Ayrton. *Photograph:* J. G. Wilson.

Concrete

can be studied, understood, and developed into a conscious design principle.

The first step towards this end is a thorough analysis of weathering in order to understand its action on various materials and finishes well enough to render it predictable. Thereafter the architect must develop his own 'philosophy of design'. Today this principle of planned and predetermined location of weathering discoloration is practised only in detail; few architects have accepted it as an over-all guiding principle governing the design of the external surfaces of the entire building.

Design and appearance

Any philosophy of weathering design must be based on the general hypothesis that rain striking the face of a building will run down taking with it any impurities that may have lodged on horizontal surfaces or be adhering to the face of the building. The tendency to assume that because concrete has properties different from those of traditional building materials it can and should be treated architecturally without regard to tradition, has tempted architects to discard many of the normal features used for the purpose of shedding water and protecting the surface. The unhappy results of these omissions have shown that just as much thought has to be given to architectural detailing with concrete as with other materials. When considering architectural detailing the following should be borne in mind.

1. Wherever a horizontal surface exists it will collect soot and other impurities from the atmosphere, these will subsequently be deposited on the face of the building by the action of wind and rain and will most likely cause streaking of the surface due mainly to the water drying off before it reaches the base of the horizontal member (Fig. 2.2). This condition can be resolved in a number of ways of which the following are but a few:

Fig.2.11. Extension to Structures Laboratory, Wexham Springs, Slough, Bucks. One means of avoiding the discoloration of concrete surfaces is to collect all the surface water into a concealed gully and to discharge it well away from the face of the structure. *Architects:* Casson, Conder & Partners. *Engineers:* Jenkins & Potter. *Photograph:* W. Duxbury.

Fig.2.13. Discoloration caused by penetration of moisture through joints in coping – lack of attention to details of this sort can lead to permanent staining of surfaces.

Fig.2.12. Bessemer Park Development, Spennymoor, Co Durham. The two types of surface finish shown in this picture, and the detailing associated with them, should ensure a satisfactory weathering of the concrete. *Architects:* Napper, Errington, Collerton, Bennett, Allott, in association with Miall Rhys-Davies. *Construction:* Bison Wall Frame. *Photograph:* Concrete Ltd.

Fig.2.14. St Anne's College, Oxford. Designed so that rain would wash the concrete surface of the building and clean it without depositing dirt. *Architects:* Howell, Killick, Partridge & Amis. *Engineers:* Harris & Sutherland. *Photograph:* Brecht-Emzig.

    a. The provision of a concealed gutter to collect the rain water and conduct it to a down pipe or spout (Fig. 2.11);

    b. By patterning the surface of the concrete, possibly the most suitable pattern being a striated finish (Fig. 2.12);

    c. By the use of an exposed aggregate facing—experience has shown that a large-size gravel aggregate is particularly effective in such situations.

2. Wherever the concentration of rainwater in a definite track can be foreseen this should be made a design feature. Typical situations where such concentrations are likely to occur are at the junctions between columns or mullions and horizontal members.

3. Where projecting copings and sills are used, discoloration of a dark colour will result on areas that are not normally washed by rain, with possibly the reverse on north-facing elevations. A change in plane or surface texture covering the anticipated depth of discoloration can help in such situations.

4. If copings are omitted from parapets and upstands they must be finished off with a steep slope away from the face to prevent the run-down of rainwater from causing streaking. Alternatively, the face of the concrete should be reeded or ribbed to discipline the flow of water over the face.

5. The design and manner in which cladding panels are used will have a distinct influence upon their over-all weathering. John Partridge (10) in describing the reasoning behind his firm's design for St Anne's College, Oxford (Fig. 2.14), writes as follows:

'St Anne's was designed so that rain would wash over the concrete surface of the building and clean it without depositing dirt—particularly sulphur-laden dirt—upon it. The elevations are highly modelled and there are no conventional

Fig.2.15. St Katherine Dock House. An excellent example of the result of discipline on the standard of architecture. *Architects:* Andrew Renton & Associates. *Engineers:* Ove Arup & Partners. *Contractors:* John Mowlem & Co Ltd. *Photograph:* Henk Snoek.

drip details oversailing the plain walls. All the drips that occur drip into wide overlapping vertical V joints, so that the water either runs down these joints or down the arrises of the concrete units themselves. *This overlapping system of concrete cladding, with its mastic joints running parallel with the face of the building, is the key to the elevational detailing.* The balcony units overlap under-window panels, and there are no 'hole in-wall' windows because these are very difficult to detail in concrete without the attendant risk of staining where water drips from the sills. Some method has to be found with hole-in-wall windows of 'designing-out' this fault—either by bringing the water in towards the building and letting it emerge again elsewhere, or by the use of modelled elements which themselves are strong enough to make the staining appear insignificant.

The buildings are finished at the top with parapets so that

there is no danger of dirty water from the flat roof being blown over onto the faces. The value of parapets in weathering has not always been fully realized because so often they have been poorly detailed. Carefully weather-proofed they are a means of ensuring a minimum water run-off onto the face of a building'.

6. Because of the shape or position of a building or parts of the same building, differential weathering can be expected wherever the following changes occur:

　　a. Direction and speed of wind;
　　b. The concentration and flow of rain water over the surface;
　　c. The aspect.

These changes should be disciplined by the use of appropriate architectural details and/or by alterations made in the nature of the surface.

The nature of the surface

Second only to architectural detailing in its influence on the weathering of concrete is the nature of the surface. Surfaces may be relatively smooth, rough, patterned, or profiled; within each of these basic categories lie a wide range of textures and finishes so that the final choice depends upon the particular designer's approach to the use of concrete.

Smooth surfaces. Generally concrete cast against smooth plain forms or moulds weathers in an unattractive manner; the reason for this is due to the skin of laitance which forms on the surface of all cast concrete. This skin is particularly vulnerable to hydration and segregation discoloration and, *inter alia*, produces a surface of variable absorbency which in its turn results in uneven weathering.

Again, when rain runs over the face of smooth concrete it takes courses determined by chance irregularities in the form of the surface and may produce light-coloured streaks on a discoloured background. It is suggested that to improve the weathering of smooth surfaces they should be acid etched or lightly abrasive blasted to remove the skin of laitance from the face; the alternative would be to pattern or profile the surface so as to mask the effects of uneven weathering.

Rough board-marked surfaces. The popularity of rough board-marked finishes is to a large extent due to the fact that surfaces of this nature provide sufficient pattern and texture within themselves to embody any blemishes that may occur on the face of the concrete. It is also in some part due to the feeling engendered by the work of Le Corbusier that rough board finishes are the *done thing*.

Exposed aggregate surfaces. Exposed aggregate finishes have been shown to have certain advantages over other smoother types of finish from the point of view of weathering. With exposed aggregate surfaces, rain running down the face of a building is broken up and distributed over a wide area so that streaking is to some extent prevented. Again, the particles of aggregate act as drips, shedding much of the water that falls on them. Another advantage of the exposed aggregate surfaces is that it can absorb moisture to much the same extent as bricks or stonework.

With walls built of dense and relatively non-absorbent materials, rain falling on the surface streams down the outer face and turns into the body of the wall at any crack or fissure and in so doing may easily reach the inner face; thus with denser or less-absorbent surfaces there may actually be greater moisture penetration, also more water may concentrate at the joints than with more absorbent surfaces.

In using exposed aggregate finishes consideration must be given to the thickness of the matrix in which stones are held and the depth of exposure of the individual stones; it is recommended that the depth of exposure should not exceed one-third the effective thickness as laid.

Fig.2.16. Surfaces consisting almost entirely of coarse aggregate particles having a smooth surface texture possess excellent weathering characteristics.
Photograph: J. G. Wilson.
Fig.2.17. Lennig House, Croydon Surrey. Spandrel beams patterned to channel water over the surface and prevent staining.
Formwork designed and supplied by William Mitchell, Design Consultants Ltd. *Architects:* Tribech, Leifer & Starkin. *Engineers:* Andrews, Kent & Stone. *Contractors:* Taylor Woodrow Construction Ltd. *Photograph:* Crispin Eurich.

Patterned and profiled surfaces. Finishes of this type are possibly the most difficult to design, and to be really successful need to be designed by an artist fully conversant not only with the properties of concrete but also experienced in the handling of the material on site.

Pattern, which implies some degree of repetition within a limited frame of reference, can be used to good effect to channel the run-off of water from the face of spandrel beams and other otherwise unbroken areas of concrete—the pattern should be conceived so that the full impact of the design becomes apparent after the surface has taken on several years of weathering. The profiling of concrete surfaces has only really become a proposition since the advent of glass-fibre reinforced polyester formwork; it can be used to enrich what would otherwise be plain surfaces and to forestall any adverse effect of weathering. Like patterned surfaces, great design skill is needed if the results are to be successful. William Mitchell, the ebullient design consultant, has perhaps done more than anyone in this, or any other, country to exploit the plasticity of concrete. Behind most of the design forms which flow from his fertile brain is the overriding principle of planned and predetermined location of weathering discoloration. Nevertheless there are situations where he is forced to accept that it is impossible to predetermine the nature of weathering; in such cases he usually resorts to a heavily modelled surface which will embrace any amount of discoloration and become an integral part of the design.

The modelled façade

One of the most exciting developments taking place at the present time is the continuing evolution of the modelled façade. It is fortunate that the idea of using one well-designed unit that could be repeated a sufficient number of times to make even the most complex of shapes a viable proposition should have appealed to some of the world's leading architects, for by so doing we have been enriched with a number of outstanding buildings. One of the earliest trend-setting examples was the Michigan Consolidated Gas Company's offices in Detroit designed by Yamasaki, who in one of his writings states 'precast concrete is the most exciting of materials—it is the future of structure'. The units used for the Consolidated Gas Company building are known locally as Detroit Gothic: they consist of precast double 'Lollipop' panels of white structural prestressed concrete having a 28 day strength of 592 N/mm^2. Each panel is 8 m high (two storeys) and when two are bolted together they frame a window in each of two storeys. The units have an exposed aggregate finish obtained by acid washing when they were seven days old.

Possibly the architects mainly responsible for pointing the way with modelled precast structural units were Skidmore, Owings, and Merrill of New York; their John Hancock buildings in St Louis and New Orleans and later the Heinz Research and Administration Centre at Hayes Park, Middlesex, and the Bank Lambert in Brussels are all well-known examples exploiting the plasticity and sculptural possibilities of precast concrete. Another architect who has used modelled concrete to make a 'positive and powerful statement' is Marcel Breuer. His design of the Research Centre at La Gaude for IBM France, a single-storey building for the Torin Coporation at Torrington, Connecticut, USA, and, more recently, the new buildings at Flaine in the French Alps are but three examples. Quite apart from these examples of modelled concrete from abroad there are a growing number in this country, all of which show in a very positive way the value of discipline imposed by the repetition of a single well-designed unit in producing an aesthetically satisfying building. One of the side benefits that can be hoped for from the use of the highly modelled façade is an improvement in the general weathering of the building; this is not to say that the building will not get just as much dirt deposited on its surface as one with a plain face, but that

Fig.2.18 (b). Close-up detail of the striated concrete finish. *Photograph:* J. G. Wilson.

Fig.2.18 (a). Elephant and Rhinoceros Pavilion, Regent's Park, London. A first-class piece of architecture produced by the synthesis of design, structure, and materials. The striated concrete finish is in perfect harmony with the function of the building.
Architects: Casson Conder & Partners. *Engineers:* Jenkins & Potter.
Contractors: John Mowlem & Co. Ltd. *Photographer:* Henk Snoek.

Fig.2.19. Czechoslovak Embassy, Kensington Palace Gardens, London. Main entrance of the Embassy building, one of the outstanding concrete buildings of the 1960s.
Architects: Jan Sramek, Jan Bocan, Karel Stepanski, in co-operation with Sir Robert Matthew, Johnson-Marshall & Partners who also acted as structural engineers. *Precast concrete cladding:* Evans Brothers (Concrete) Ltd. *Photograph:* Richard Einzig.

Fig.2.20. Mural on Board of Trade Building, Victoria Street, London, SW 1, designed by William Mitchell Design Consultants Ltd. Photograph taken shortly after completion.
Photograph: J. G. Wilson.

the movement and run-off of rainwater over and from the surface will take place in such a manner that it will enhance rather than detract from the general appearance of the building. This calls for a careful study of profiles to establish those which are good, and those which are not so good.

Silicones

It was not until 1953 that silicones were manufactured commercially in this country, and today although there are a great number of 'silicones' produced under different trade names, the raw materials on which these are based come mainly from only two or three suppliers. Unfortunately the claims made for many of these materials have, so far as concrete is concerned, to say the least of it, been disappointing.

Wilson (11) has stated:

Silicones have the effect of imparting water-repellent properties to the surface pores so that instead of holding water they shed it. Besides preventing the absorption of water it appears reasonable to assume that a silicone treatment may assist in keeping the concrete clean by preventing the absorption of soot or grime into the surface. Even if it does not assist in keeping the surface clean—there is still a difference of opinion on this score—its use will make any subsequent cleaning easier.

Figs.2.21 (*a*), (*b*). Two examples of exposed aggregate concrete after one and a half years' exposure in a moderately polluted atmosphere. The coarse aggregate particles in the untreated panel have remained reasonably clean, whereas those in the panel treated with a Silicone have become coated with grime.

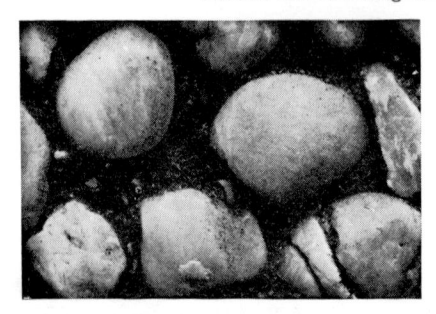

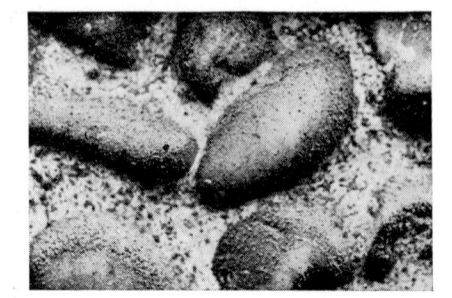

Since this was written in 1962, evidence has been produced that some water-repellents can, when applied to certain types of exposed aggregate finishes, cause the aggregate particles to become coated with a sticky film which never really hardens and to which impurities readily adhere (Fig. 2.21). It has also been shown that many of the so-called silicone water-repellents lose their globulation properties in a matter of months after application. As for keeping the surface clean, or the claim that their use will make subsequent cleaning easier, evidence suggests that these claims are only supportable over a short period of time. The effectiveness of any silicone-based water-repellent is dependent to a very large extent on the nature and quality of the concrete to which it is applied. Reference should be made to BS 3826.

Clear coatings

Clear coatings may be applied to a variety of concrete surfaces for a number of reasons. In the case of exposed aggregate finishes it is sometimes found that the aggregate chosen for the facing appears dull and lifeless when dry, but strong to lively when wet; to maintain the wet appearance the surface can be given a clear coating.

Other conditions calling for the possible use of a clear coating are:

1. Where the building is situated in a chemically polluted atmosphere;

2. Where white cement concrete has been used and it is desired to prevent the surface from discoloration due to atmospheric contaminants;

3. To avoid the change in colour which takes place on grey cement concrete when the surface is wetted by rain;

4. To prevent soiling of surfaces in positions where concrete is exposed in pedestrian areas;

5. To facilitate subsequent cleaning of the surfaces.

Virtually no research appears to have been carried out in this country to establish which of the clear coatings available on the market more nearly fulfil the above requirements. In the USA the Portland Cement Association Research and Development Laboratories have carried out an investigation to establish the effectiveness of a variety of coating materials in protecting white concrete against discoloration from atmospheric contaminants. The following is an extract from their report (12):

'Sixty clear coatings were investigated for their ability to protect the surface and maintain the original color of exposed-aggregate and smooth white concrete. Coated concrete specimens were subjected to 30 cycles of an accelerated weathering test which included exposure to dilute sulfurous acid and salt solutions, fly ash, ultraviolet and infrared rays, and freezing and thawing. At the conclusion of these tests the appearance of the specimens ranged from practically no change to deep shades of grey and brown.

Analysis of the coating materials by infrared spectroscopy indicated that many of the better coatings on the exposed-aggregate surfaces consisted mainly of the methyl methacrylate form of acrylic resin. While minor additions of ethyl acrylate and butyl methacrylate were present in some coatings, those based mainly on these materials were less satisfactory. A single sample of material identified as vinyl toluene-acrylic copolymer also rated among the better coatings.

The test ratings of these coatings when used on smooth concrete surfaces indicated that the low-viscosity, low-solids-content materials did not generally provide sufficient film thickness to adequately protect the surface paste from soiling. Certain higher viscosity resin coatings protected the surface more adequately; however, these were not as satisfactory as thicker coatings based on methyl methacrylate.

A limited number of coatings representing good, average and poor performance in the accelerated weathering test, were used on exposed-aggregate specimens placed in an outdoor plot in a highly industrial area. The results of these tests tended to verify the results of the laboratory tests.

The tests reported here indicate that properly selected clear coatings protect and maintain the appearance of white concrete exposed to atmospheric contaminants present in industrial areas. Such proper coatings permit relatively easy cleaning of surface dirt. The use of many of the coating materials included in this study resulted in permanent discoloration of the concrete surface. In areas of little or no air pollution it may be advisable not to use coatings on architectural concrete.'

Cleaning concrete surfaces

It is not necessary to describe in detail the methods which can be used for cleaning the external surfaces of buildings constructed of, or faced with, concrete; these are identical to those used on natural stone or brickwork, details of which are given in BRS Digest no. 113 (13).

In recent times the use of abrasive blasting, either dry or wet has, to some extent, superseded the more normal method of cleaning by means of water spray—either of these methods are fast compared with the normal method, or that of steam cleaning. On the other hand, abrasive blasting entails the risk of damaging the surface and causing a noise, and in the case of dry abrasive blasting, a dust nuisance to neighbours. Whatever method is used proper safety precautions must be taken. Advice on any safety aspect can be obtained from HM Factory Inspectorate, Baynards House, 1–3 Chepstow Place, Westbourne Grove, London, W 2.

REFERENCES

1. BUILDING RESEARCH STATION (1963). 'Shrinkage of natural aggregates in concrete', *BRS Digest*, no. 35 (Second Series) (London: HMSO).
2. BRITISH STANDARDS INSTITUTION (1969). *Code of Practice for the structural use of concrete* (in draft).
3. BUILDING RESEARCH STATION (1968). 'Concrete in sulphate-bearing soils and ground water', *BRS Digest*, no. 90 (Second Series) (London: HMSO).
4. CEMENT AND CONCRETE ASSOCIATION (1967). *Recommendations for the production of high quality concrete surfaces* (London: C & CA).

5. Murphy, W. E. (1964). 'Some influences of concrete mix design and method of placing on the surface appearance of concrete', Paper presented to Symposium on *Surface Treatment of In-situ Concrete*, London, 28 September 1964 (London: Cement and Concrete Association).
6. Levitt, M. (1963). *Consolidated report on the crazing of cast stone* (1930–1962) (London: Cast Stone and Cast Concrete Products Industry).
7. Blake, L. S., Kinnear, R. G., and Murphy, W. E. (1964). 'Recent research into factors affecting the apperance of in-situ concrete', Paper presented to Symposium on *Surface Treatment of In-situ Concrete*, London, 28 September 1964 (London: Cement and Concrete Association).
8. Building Research Station (1952). 'The control of lichens, moulds and similar growths on building materials', *BRS Digest*, no. 47 (First Series) (London: HMSO).
9. Wilson, J. G. (1969). *White concrete with some notes on black concrete.* (London: C & CA).
10. Cement and Concrete Association (1969). *Concrete Quarterly*, no. 82 (July–September).
11. Wilson, J. G. (1962). 'Finishes to in situ concrete', *Exposed concrete finishes*, vol. 1, p. 26 (London: C. R. Books).
12. Litvin, A. (1968). 'Clear coatings for exposed architectural concrete', *J. Portland Cement Assoc. Res. and Development Lab.*, **10** (2), 49–57 (May), Portland Cement Association Development Department Bulletin D 137.
13. Building Research Station (1969). 'Cleaning external surfaces of buildings', *BRS Digest*, no. 113 (Second Series) (London: HMSO).

APPENDIX 1

List of British Standards and British Standard Codes of Practice referred to in this chapter.
BS 12 (1958) *Portland cement (ordinary and rapid-hardening).*
BS 146 (1958) *Portland blast furnace cement.*
BS 4027 (1966) *Sulphate resisting Portland cement.*
BS 915 (1947) *High alumina cement.*
BS 882 (1965) *Coarse and fine aggregates from natural sources.*
BS 1047 (1952) *Air-cooled blast furnace slag coarse aggregate for concrete.*
BS 877 (1967) *Foamed blast furnace slag for concrete aggregate.*
BS 3797 (1964) *Lightweight aggregates for concrete.*
BS 3826 (1967) *Silicone-based water repellents for masonry.*
CP 114 (1969) *The structural use of reinforced concrete in buildings.*
CP 116 (1965) *The structural use of precast concrete.*
In draft, 'Code of Practice for *The structural use of concrete.*'

APPENDIX 2

SPECIFICATION CLAUSES COVERING THE PRODUCTION OF HIGH-QUALITY FINISHES TO *IN SITU* CONCRETE

Note. This Appendix is an abridged version of the specification bearing the same title obtainable from the Cement and Concrete Association, 52 Grosvenor Gardens, London, SW1.

Scope

These specification clauses are for use where *in situ* concrete is to be the face material of a building or structure and the finish on the concrete is of first importance. The clauses do not of themselves constitute a complete specification. They are for use where the finish is to be of a high quality and clauses in addition to those contained in the relevant BS Codes of Practice, or which differ from those for ordinary concrete work are required. Where explanatory information is considered desirable this is included and follows the clauses to which the information applies. The clauses refer only to the use of timber formwork. Other materials such as steel, plastics, glass-fibre reinforced plastics, rubber, etc., are used for formwork or form-linings and are capable of producing high-quality finishes. Their use will require clauses related to their requirements which are outside the scope of this specification.

Introduction

The production of finishes on cast *in situ* concrete to the high standards which are sometimes required necessitates providing contractors with a well-written specification in order that they may clearly understand what the requirements are. Unnecessarily exacting requirements may make contractors question the specification as a whole or may render it virtually impossible for them to be sure just what is expected. They may be over-cautious and over-price, or not cautious enough and under-price. Nevertheless, the designer cannot properly expect his intentions to be fully carried out unless the requirements are clearly specified.

The degree of refinement that can be achieved with concrete must, of course, be appreciated and it is the duty of the designer to make himself fully conversant with the practical limitations of the material and design accordingly. The specification should include and cover such items as the number, location and details of all construction, contraction, and expansion joints that will be required or permitted and details of any special features required in connection with the finish to the concrete. Having satisfied himself that the contractor fully appreciates all the implications of the contract documents it is then up to the designer to see that he obtains the standard of finish he requires.

On such finishes freedom from defects such as honeycombing, spalling, damage to the surface, and cracking which would spoil the appearance is essential. Requirements for reinforcement to resist stresses due to dead and live loads are generally well understood and carefully observed, but the necessity to design for movement is not perhaps always realized. To avoid the risk of shrinkage cracking the reinforcement should be adequate and be properly distributed to resist stresses due to volume changes caused by both temperature and moisture variations. Information on this subject is contained in *Movement joints in concrete* (1).

SECTION ONE: DEFINITIONS

1.1 *Fair-faced concrete*
Concrete cast against plywood, wrought timber, steel, or other smooth face material which is put together in such a manner that there is no leakage of grout between boards or formwork panels or at corners, and so that the pattern of the boards or panels may form an inconspicuous but important feature. The concrete to be dense, and as far as possible of uniform colour without blow-holes, fins, or other surface blemishes. Construction joints to be sensibly arranged and either made as inconspicuous as possible or given deliberate emphasis. Apart from filling bolt-holes, the surface of the concrete to be left as struck from the formwork.

1.2 *Sawn or rough board finish*
The surface of concrete cast against selected timber boarding having a pronounced grain pattern, or boarding that has been sawn or abrasive blasted to provide a textured surface, the formwork to be put together in such a manner that there is no leakage of grout between boards or formwork panels or at corners, and so that the pattern of the boards or panels forms an important feature. The concrete to be dense, and so far as

Appendix 2 87

possible of uniform colour without blow-holes or other surface blemishes. Apart from filling bolt-holes, the surface of the concrete to be left as struck from the formwork.

1.3 *Formwork*   A temporary structure built to contain fresh concrete so as to form it to the required shape and dimensions and to support it until it hardens sufficiently to become self-supporting. Formwork includes the surface in contact with the concrete and the necessary supporting structure.

SECTION TWO: MATERIALS

2.1 *General*   All the materials shall comply with the appropriate British Standard unless otherwise agreed in writing.

DISCUSSION. *It is not possible to know precisely which of the various British Standards (see Section 5D) the reader of this publication will be using. He should, however, refer individually to the relevant British Standards in his specification.*

2.2 *Supply of cement*   Where uniformity of finish is required no change in the type of cement shall be permitted during the course of construction and to ensure as far as possible that the colour of the concrete is constant throughout the contract the cement shall come from one works.

DISCUSSION. *Sulphate-resisting Portland cement is normally slightly darker in colour than most brands of ordinary or rapid-hardening Portland cement.*

2.3 *Supply of aggregates*   The contractor shall make arrangements for the designer or his representative to visit the pit or quarry to inspect and approve the materials and shall establish that ample supplies of both fine and coarse aggregates of the quality and colours selected are available to complete the contract.

    The coarse aggregate shall be obtained in at least two grades (or as many as the engineer or architect may specify). They shall be stored on a clean hard base in separate compartments or in approved hoppers. Suitable precautions shall

G

be taken to prevent mixing of the sizes before they are batched. The storage area shall be as clear as possible of trees and care shall be taken to avoid contamination by leaves and any other organic material from whatever source.

DISCUSSION. *All natural aggregates should preferably comply with the requirements of BS 882 'Coarse and fine aggregates from natural sources', but special examination should be made to ensure that the aggregate is free from impurities such as pyrites and coal which would stain the surface of the concrete. This may necessitate making test samples incorporating suspect aggregate or sand. The aggregate must remain uniform in colour during the course of the contract. Uniformity in colour of the fine aggregate is of special importance; colour of the coarse aggregate is only important when the aggregate is to be exposed.*

The grading and particle shape of the aggregates must be kept sensibly constant—variations in either respect of the fine or of the coarse material could make adjustment of the mix proportions necessary in order to keep the workability of the concrete constant. This will result in changes in colour on the finished work, or where the aggregate is continuously graded and is being exposed (see Sections 5B and 5C) in an uneven distribution of the aggregate in the surface.

Stockpiling of the coarse aggregate in at least two grades makes closer control of the over-all grading possible and is especially desirable when the materials are 'sandwich-loaded' on the delivery vehicles by the supplier. Separate delivery and storage, for example of the 10–5 mm ($\frac{3}{8}$–$\frac{3}{16}$ in) size material of a continuously graded 20-mm ($\frac{3}{4}$ in) coarse aggregate from the 20–10 mm ($\frac{3}{4}$–$\frac{3}{8}$ in) size and the separate batching of these sizes will contribute to the maintenance of constant grading.

The fine aggregate should preferably be natural sand within grading Zones 2 or 3. Zone 1 sands are not recommended because their lack of fines results in more blow-holes and colour variations; Zone 4 sands may cause 'aggregate transparency', even in a mix that is well designed and controlled.

For fair-faced and board-marked finishes the aggregates as a rule should be continuously graded. On the other hand, for all heavily exposed aggregate finishes gap-graded aggregates, in which the coarse particles of sand and finer particles of coarse aggregate are omitted, give the most satisfactory results (see Section 5C).

Apart from the aggregates covered in BS 882, there are others which, although not strictly complying in all respects with this Standard, will nevertheless produce a first-class concrete when used in accordance with good concreting practice. One such aggregate is calcined flint.

Appendix 2 89

2.4 *Ready-mixed concrete* — When ready-mixed concrete is used suppliers must guarantee that the cement and aggregates used in the mix comply in all respects with Clauses 2.2 and 2.3 and the concrete mix design with the requirements of 3.1

2.5 *Release agents* — The release agent shall be a material marketed as such and shall be of one of the following types:

1. Cream emulsion;
2. Neat oil with surfactant added;
3. Chemical release agent.

The brand used shall be approved in writing. Cream emulsion shall not be used in conditions where they may be subject to freezing.

Release agents shall be stored strictly in accordance with the manufacturers' instructions; if a cream emulsion is used, particular care shall be taken to protect it from extremes of temperature. The method of storage shall be such that contamination cannot occur and the release agent must not, at any time, be diluted by the addition of water or other materials.

SECTION THREE: CONCRETE

3.1 *Concrete mix design* — The concrete mix shall be designed in accordance with the relevant Code of Practice or other structural requirements and in addition the cement content shall be not less than the values given in the table below.

Minimum cement content for mixes to give good appearance.

Nominal maximum size of aggregate. mm (in)	Minimum cement content of fully compacted concrete kg/m^3 (lb/yd^3)
40 $(1\frac{1}{2})$	330 (560)
20 $(\frac{3}{4})$	370 (620)
10 $(\frac{3}{8})$	420 (710)

Standard mixes may be permitted provided the same standards of quality control and testing as for designed mixes are used.

The contractor shall obtain samples of the materials to be used and shall prepare trial mixes from these materials to confirm the suitability of the concrete mixes in both the plastic and hardened state.

DISCUSSION. *In selecting the concrete mix, care must be taken to see that the selection of materials and the mix proportions are related to the particular finish required and not only to strength and durability considerations. In particular, this involves ensuring that the workability of the concrete is appropriate to the means of compaction and that a fully compacted dense concrete can be produced throughout the site construction.*

3.2 *Sample panels*

In addition to the trial mixes required—Clause 3.1—the contractor shall allow for providing as early as possible in the contract period a sample panel at least 2 m (2 yd) square in accordance with the specification to confirm the suitability of the chosen mix and the formwork to give the desired finish. The sample shall include one horizontal construction joint to the details shown on the drawings and shall be finished exactly in accordance with the specified finish.

A second similar sample panel is also to be allowed but will only be required if it is necessary to consider detail variations to the specification to obtain the quality of finish required. (Where more than one type of finish is required sample panels of each type shall be provided.)

The architect or the engineer may call for adjustments to the materials and the mix proportions after inspection of the sample panels. No adjustments to the agreed mix shall be made without the consent of the engineer and, where applicable, the architect. The accepted panel is not to be destroyed until the contractor is instructed to do so by the architect.

3.3 *Placing and compacting*

The concrete shall be placed in one continuous operation rising uniformly in the formwork at a rate exceeding 2 m/h (6·6 ft/h). The concrete shall not be handled in any manner that may encourage it to segregate, for example, it shall not be allowed to fall freely down an incline nor shall it be placed in concial heaps. If the concrete is allowed to fall freely through a height of more than 2 m (6·6 ft) precautions shall be taken to ensure that neither the formwork nor the reinforcement are damaged or displaced.

Appendix 2

The concrete shall not be placed directly against a vertical form face but shall be caused to flow to this surface during the compaction process. Care shall be taken to avoid the form face being splashed with mortar during the placing operation.

The concrete shall be compacted by immersion vibrators having a frequency of at least 8000 cycles/min; compaction shall start as soon as there is sufficient concrete within the formwork to immerse the vibrator and vibration shall continue during the placing operation so that at no time shall there be a large volume of uncompacted concrete in the formwork. Sufficient vibrators shall be available to ensure that delays, or incomplete compaction, are not caused by mechanical breakdowns.

Where, one hour after completing a lift, bleeding of the mixing water is apparent, external vibration shall be applied to the top 300 mm (12 in) of the concrete and if, because of this additional vibration, the concrete settles in the formwork, it shall be topped up with fresh concrete of the same mix and tamped level with the top of the formwork.

DISCUSSION. *The suitability of the concrete for placing should always be judged when the concrete is at the point of placing. Workability tests should be carried out in a position close to the work in hand but away from the effects of any vibration. These tests should be made on a representative sample of the concrete at the beginning of concreting operations and whenever a gross change of section or method of compaction is used, and at more frequent intervals if the engineer's or architect's representative considers them necessary.*

When concreting is delayed for any reason the concrete already in the formwork should be compacted and the vibrators stopped and withdrawn.

Differences in appearance caused by delays in placing may be judged sufficient grounds for condemning the finished concrete.

3.4 *Construction joints*

Construction joints shall be prepared in accordance with the instructions of the designer. Where no special instructions are given horizontal joint surfaces shall be well brushed with a wet banister brush and washed off with a spray of water two to four hours after casting to expose the aggregate and to provide a key for the next lift. Before the formwork for the next lift is placed, the exposed aggregate surface should be well scrubbed to remove all laitance and loose material. No mortar shall be placed on the treated surface before placing the next lift of concrete.

3.5 Cold weather concreting

No concrete shall be placed on ground which is frozen or in formwork which has a layer of frost or ice on the surface or around reinforcement which is coated with frost or ice. The temperature of the concrete when placed in the form shall be at least 10 °C (50 °F).

Insulation shall be provided for fresh concrete when there is any likelihood of attack by frost. The insulation shall be continuous over all surfaces exposed to possible damage and shall be firmly fixed in place. It shall remain in position for at least two days where night frosts only are involved and for such longer period as may be necessary, especially when average air temperatures are continuously below 2 °C (35 °F).

DISCUSSION. *Before any concrete is placed in cold weather the contractor should satisfy the engineer's or architect's representative that sufficient materials and equipment such as insulation, heaters, covers, etc., are available on site to ensure that the specified temperature can be maintained.*

It is more difficult to obtain good concrete surface finishes when the concrete temperature is low than when it is higher. For this reason the specified minimum temperature of the concrete when placed is higher than that required by normal structural requirements when the minimum temperature is normally specified as 5 °C (41 °F.).

Guidance on cold-weather concreting is contained in Winter Concreting *(2).*

3.6 Thermal shock

Precautions shall be taken to insulate or protect the concrete against thermal shock caused by exposing warm concrete surfaces to low temperatures on removal of formwork and other insulation.

DISCUSSION. *During the hydration of cement a considerable amount of heat is evolved which raises the temperature of the concrete in the forms. Insulation applied to the concrete surfaces as provided by the forms can create a large temperature differential between the concrete surface and the surrounding air. Sudden lowering of the surface temperature by exposure to cold air or curing water can lead to cracking and crazing of the concrete.*

3.7 Rust staining

All starter bars are to be covered with plastic sleeves or, where heavy reinforcement projects through the concrete, with a waterproof cover. Alternatively, projecting steel shall be

Appendix 2   93

thoroughly but thinly coated with a cement–polyvinylacetate grout.

DISCUSSION. *One of the chief causes of rust staining is the use of steel scaffolding; on all contracts where the finish of the concrete is of paramount importance non-ferrous scaffolding should be used or the scaffolding be efficiently painted.*

Provided the reinforcement is securely fixed with the amount of cover recommended by the relevant Code of Practice, and provided that the concrete is fully compacted, staining by rusting of the reinforcement should not occur once the concrete has been cast. However, staining is sometimes caused by wire ties protruding too near the surface or off-cuts being left on soffit formwork. Staining sometimes results from rust scale falling or being washed by rain on to the form.

Because contamination of this type is most likely to occur with soffits, it is recommended that special attention be given to thorough cleaning of soffit forms by compressed air lines immediately before concreting. Cleaning with a water hose cannot be recommended because of the danger of water ponding on the forms and of mould oil being removed.

3.8 *Making good*

Making good shall be kept to the absolute minimum. Where it is essential, for example, filling tie-rod holes, it is to be done using the same mix as the concrete, except that the coarse aggregate shall be excluded and that the cement shall consist of a mixture of white and ordinary grey Portland cement contained in such proportions that the colour of the making-good mortar matches exactly that of the surrounding hardened concrete. The area is to be cleaned thoroughly first by brushing and washing with clean water and the filling is to be well compacted by trowel or rod and finished with a sponge-rubber-faced float.

Great care is to be taken in making good so as not to spread the operation beyond the holes.

DISCUSSION. *When mortar with exactly the same proportions as that in the body of the concrete is used for making good, this invariably results in a mortar of a darker colour. This difference in colour is due to the different curing conditions, to the effect of the form face material on the surrounding concrete, and to floating or trowelling which causes moisture movements, sometimes bleeding, from the filling mortar.*

A close colour match can be obtained by using a mortar prepared to the same proportions as that of the concrete, provided that the mortar is placed against a similar form face and cured in the same way as the concrete. In the case of tie-bolt holes this can be done by fixing a plate of appropriate material to the face of the concrete before filling the hole and leaving the plate in position for the required curing period. However, an alternative method and the only one which can be used when filling blow-holes and other similar surface defects of grey concrete, is to lighten the colour of the filling mortar by the use of white cement. In this case the mix proportions of the mortar should be the same as that in the concrete (for example, for a 5 : 1 concrete mix with 40 per cent of fine aggregate, the sand/cement ratio of the mortar should be 2 : 1). The cement used should be composed of about 3 parts white cement to 1 part grey, but it is advisable to make some test pieces during the course of the contract to establish the most suitable proportion of white/grey cement. The colours of the test specimens should not be judged until after 28 days. The workability of the mortar should be sufficient to allow full compaction by rodding.

Before filling, holes must be cleaned thoroughly by brushing and washing with clean water to remove loose particles and dirt; when the part to be filled has dried sufficiently, so that the surface is just damp, then the filling mortar must be well rammed and compacted by trowel or rod and then finished by a sponge-rubber-faced float. Finishing with a steel trowel or knife is not recommended because this produces too dark a colour.

With the approval of the designer, tapered plastic plugs may be used in place of cement mortar for the filling of tie-rod holes.

SECTION FOUR: FORMWORK

4.1 *General* Before construction of the formwork is started, the contractor shall submit to the engineer and architect his proposals for the design, use, and reuse of formwork.

Notwithstanding any approval which may be given, it is the contractor's responsibility to ensure the adequate performance of the formwork in relation to the safety of site personnel and to ensure that the finished concrete produced by the formwork is in accordance with this specification.

DISCUSSION. *The design, construction, and use of formwork, in common with other temporary works, is generally the responsibility of the contractor and in most cases it will be inappropriate for the engineer and architect to study the contractor's proposals in great detail, but they should satisfy themselves on the following general points:*

Appendix 2 95

    a. *the proposed sheeting material and release agent will produce the specified concrete surface finish;*

    b. *The general design and stiffness of the formwork will be capable of producing concrete within the specified tolerances;*

    c. *The formwork will be capable of maintaining the specified standards to the end of the proposed number of reuses.*

More detailed guidance on specific requirements is given under the appropriate clauses in this specification.

4.2 *Design of formwork*

The design of the formwork shall be the responsibility of the contractor unless otherwise agreed. Formwork shall be designed by competent personnel experienced in the design, construction, and use of formwork for the production of high-quality concrete surface finishes. The design of the formwork shall be such as to produce concrete which, in every respect, will be in accordance with the drawings and specification. The design shall also be such that the formwork may be erected, supported, braced, and maintained so that it will safely support all loads that might be applied until the concrete has hardened sufficiently to support such loads itself. The formwork must be capable of being struck and removed safely and without causing shock or excessive stress being applied to the concrete structure.

    DISCUSSION. *Attention is drawn to the importance of the design of the formwork which will have much to do with the over-all economy and quality of the finished work since the cost of formwork is between 35 and 60 per cent of the total cost of the concrete structure.*

    Among the points of importance to the contractor when designing his formwork, and when pricing and programming his work are:

    a. *The number of reuses he can obtain from his forms;*

    b. *The position and treatment of joints in the concrete and the form sheeting;*

    c. *Whether form-ties will be allowed and if so their positions and the treatment of the holes left in the concrete;*

    d. *The type and standard of surface finish required;*

    e. *The tolerances allowed.*

    The engineer and architect should have some knowledge of how these items affect the contractor's work and prices and should bear them in mind when designing the permanent structure. The requirements for points b *and* e *should be clearly stated in the contract documents, but the best results will be obtained if all parties collaborate in deciding the final details before work starts.*

Extra provisions are essential for formwork for special structures such as shells and folded plates and for special construction methods. The basic geometry of such structures, as well as the required camber, must be given in sufficient detail to permit the contractor to design the formwork. Where a particular finish is required on the face of the concrete this must be fully described and where possible illustrated.

4.3 *Inspection and checking of formwork*

Before any concrete is placed the formwork shall be inspected and passed by the engineer or architect. Notwithstanding such passing of the formwork, it will remain the contractor's responsibility to ensure the safety of the work and that the finished concrete complies with the specification.

DISCUSSION. *In order to obtain a high-quality concrete surface finish, the inspector should check the following:*
  a. *The forms will produce concrete within the required dimensional tolerances;*
  b. *The interior of the form is free from all foreign matter;*
  c. *The correct form sheeting material has been used and this is clean;*
  d. *The release agent has been thinly and evenly applied without build-up or puddles anywhere;*
  e. *All form joints are clamped tight and sealed against loss of grout from the concrete.*

4.4 *Release agents*

The release agent shall be applied evenly in a very thin film. The type of release agent should be one recommended by the manufacturers for use with the type of form face and under the conditions proposed. The manufacturer's recommendations on rate of spread and method of application shall be obtained and followed.

The release agent shall, wherever possible, be applied the day before concreting. Prior to their first use, absorbent materials, such as raw timber or sanded plywood, shall be given three applications of release agent at intervals of at least a day.

DISCUSSION. *Only rarely will the use of any proprietary release agent cause staining of the concrete surface, unless it contains impurities or an excessive quantity is applied. Some release agents do, however, en-*

Appendix 2

courage blow-holes; others affect the density and durability of the surface and hence cause colour variations. The three types of release agents listed in Clause 2.5 of this specification are those most likely to inhibit the occurrence of blemishes. The optimum rate of spread of release agent will depend upon the type of release agent used and the nature of the form face.

It is generally true that the optimum quantity of release agent is as little as possible, providing only that the whole of the form face is evenly coated; considerable care must therefore be exercised in the application of the release agent.

4.5 Prevention of leakage

So as to prevent leakage or escape of grout between individual boards, they shall be grooved on both edges and fitted with loose hardboard tongues. Alternatively a strip of foamed polyurethane shall be cramped up tight between individual boards.

Corner pieces are to be formed out of the solid to produce 20 mm ($\frac{3}{4}$ in) rounded external and internal angles in the finished work and are to be tongued to the boarding. To prevent leakage of grout between panels and between subsequent lifts of concrete a strip of flexible foamed polyurethane (25 mm × 10 mm (1 in × $\frac{3}{8}$ in) is a convenient size) shall be stuck or pinned to the edge of each panel and to the face of the forms used for subsequent lifts.

4.6 Striking of formwork

Formwork shall be struck without shock or vibration which might damage the concrete (see also Clause 3.5).

In no circumstances shall the formwork be struck until the concrete has reached a cube strength of at least twice the compressive stress to which the concrete may be subjected at the time of striking.

When forms are struck there must be no excessive deflexion or distortion and no evidence of damage to the concrete, caused either by removal of support or by the striking operation.

Supporting forms and props must not be removed from beams, floors, and walls until these structural units are strong enough to carry their own weight plus any superimposed load, which at no time should exceed the live load for which the unit was designed.

DISCUSSION. *The formwork to vertical surfaces such as beam sides, walls, and columns may be removed in special circumstances after 16–18 hours in normal weather conditions, although care must be exercised to avoid damage to the concrete, especially to arrises and other features. In cold weather a longer period may be necessary before striking, and protection against frost damage (Clause 3.5) should be afforded to the concrete. In cases where the early concrete strengths are not accurately known and where very little live load will be applied to the structure before the specified 28 day strength has been reached, the following minimum times (in days) shall elapse before removal of formwork:*

Recommended minimum times for striking formwork when the member is carrying only its own weight.

	Ordinary and sulphate-resisting Portland cement concrete		Rapid-hardening Portland cement concrete	
	Cold weather	Normal weather	Cold weather	Normal weather
Air temperatures	About 3 °C (37 °F)	About 16 °C (60 °F)	About 3 °C (37 °F)	About 16 °C (60 °F)
	Days	Days	Days	Days
Slabs (props left under)	7	3½	4	2
Beams soffits (props left under)	14	7	8	4
Props to slabs	14	7	8	4
Props to beams	28	14	16	8

Notes

1. All days during which the average temperature is below 2°C (36°F) shall be added to the minimum times given in the table.

2. The times given for removal of props are based on the dead weight to be supported being not more than about half the total design load. For horizontal members designed for very light live loads, these times may need increasing.

SECTION FIVE

A. ADDITIONAL SPECIFICATION CLAUSES FOR A SAWN OR ROUGH-BOARD FINISH

Selection of boards

The boards for this finish shall be either boards selected from well-weathered or otherwise prepared stock having one face with a raised and prominent growth-ring figure or boards

Appendix 2 99

wrought on one side and two edges with the other side left sawn.

Before making the formwork, the contractor shall submit samples of commercial softwood boards to the designer for him to select the texture of board required. The sample boards are to be representative of the timber proposed for the form sheeting and normally readily available from the timber mills for the contract period. Commercial board widths shall be used and the sizes of the boards shall be agreed when the samples are submitted.

The selection and approval of the boards shall not relieve the contractor from responsibility for ensuring an adequate supply of the required quality to produce the desired finish throughout the work.

Preparation of formwork

In preparing the formwork, boards with a strong texture shall be mixed with boards having a less-pronounced texture and not grouped together.

Boards are to be of one length, or stopped ends of adjoining boards are to be staggered 600 mm (24 in), the joints being carefully aligned to a pattern approved by the designer.

Thickness of boards

The thickness of the timber shall be 32 mm ($1\frac{1}{4}$ in). A tolerance of 0·8 mm ($\frac{1}{32}$ in) in the thickness of the boards will be accepted, and a proportion of the boards should show this variation which no attempt should be made to conceal.

Alternatives

a. The contractor, with the approval of the designer, may in place of using 32 mm ($1\frac{1}{4}$ in) thick boards use 25 mm (1 in) boards providing the stud spacing is reduced and the requirements of Clauses 4.1 and 4.2 are fully complied with.

b. The contractor, with the approval of the designer, may in place of using 32 mm ($1\frac{1}{4}$ in) thick boards attached to studs, use 13 mm ($\frac{1}{2}$ in nominal) thick boards attached to a backing of plywood in which case the boards need not be grooved and fitted with loose tongues—but in all other respects they shall be as specified for 32 mm ($1\frac{1}{4}$ in) thick boards.

Width of boards	The boards shall be of equal widths in relation to the length or depth of the section to be cast or as otherwise specified by the designer.
Assembly of form panels	The boards are to be securely nailed to the studs to an even approved pattern, the nail heads being driven flush with the surface and not punched in. The form panels shall be supported by members of sufficient dimensions and at sufficiently close centres to prevent deflexion of the panels during the placing and compaction of the concrete, which would produce finished concrete outside the specified tolerances. *Note.* It is suggested that the form panels should be well wetted down before placing the concrete to close all joints and to prevent expansion of the form sheeting.
Re-use of forms	No panels or boards are to be used more than six times unless specifically approved by the designer.

B. ADDITIONAL SPECIFICATION CLAUSES FOR A BUSH-HAMMERED OR POINT-TOOLED FINISH

Concrete mix	The concrete mix shall be based upon a mimimum cement content of 360 kg/m³ (600 lb/yd³). See also Clauses 3.1 and 3.2.
Cover to reinforcement	For all exposed external work the cover required by the relevant Code of Practice shall be increased by 5 mm (0·2 in) to allow for material removed by tooling.
Age of concrete before tooling	Concrete shall not be tooled until it has reached a compressive strength of 25 N/mm² (3600 lbf/in²) without the approval of the designer.
Arrises	Tooling shall stop short of all arrises unless otherwise directed by the designer.

Appendix 2 101

Bolt-holes   Unless otherwise detailed, immediately upon removal of the formwork the tie-rod holes shall be made good in accordance with Clause 3.8.

DISCUSSION. *The purpose of tooling is to remove the skin of hardened concrete paste from the face of the concrete. The tools used for the purpose vary from hand hammers used alone or with chisels to a wide range of electric and compressed-air tools. The most commonly used tool is the bush-hammer; this, when electrically operated, normally consists of either a circular head with twenty-one cutting points or a roller head with ninety cutting points. Tools operated by compressed air consist mainly of cruciform bits used singly or in groups of two or more. These tools are generally known as scaling hammers. Apart from these there are other more sophisticated tools which have been developed, mainly with the object of reducing the time taken to remove the surface skin.*

Point-tooling is a finish produced by the use of a single point tool. If a light texture is required the tool should have a short drawn point; if a heavy texture is required the tool should have a long drawn point.

It is frequently thought that by tooling the face of concrete, blemishes caused by leaking formwork, blow-holes, and honey-combing caused by inadequate compaction, etc., will be masked. This is not altogether correct. Bush-hammering will quite frequently reveal imperfections in the concrete which were not previously visible. It will also emphasize such items as badly formed lift lines and blow-holes. On the other hand, point-tooling, because of the coarse texture it gives to the concrete, can be used to good effect for the purpose of masking many common blemishes.

It is vitally important to ensure that grout-tight forms are used where concrete is to be tooled and with this end in view it is advisable to tape all panel joints on the inside face to prevent leakage.

If concrete is tooled before it has reached a minimum compressive strength of 25 N/mm^2 (3600 lbf/in^2) there is always the danger that instead of the aggregate particles merely being cut or bruised they will be forced out of the surface of the concrete. An exception to this rule can be made when using limestone aggregate concrete, and tooling by means of a 'Jason pistol'.

With all forms of tooling it is essential to stop the tooling short of arrises; if this is not done spalling will invariably result. The alternative is to chamfer or round the arris, in which case the tooling, if carefully done by hand when the concrete is at least three months old, can be taken around the arris without much fear of damage.

C. ADDITIONAL SPECIFICATION CLAUSES FOR FINISHES OBTAINED BY ABRASIVE BLASTING

Concrete mix — The concrete mix shall be based upon a minimum cement content of 330 kg/m³ (560 lb/yd³). See also Clauses 3.1 and 3.2.

Compressor — The air compressor used in conjunction with abrasive blasting shall be capable of producing 2·83m³ (100 ft³) of air per minute and providing a pressure of 0·83 N/mm² (120 lbf/in²) at the nozzle.

Bolt-holes — Unless otherwise detailed, immediately upon removal of the formwork the tie-rod holes shall be made good in accordance with Clause 3.8.

DISCUSSION. *The concrete mix should be designed to provide the finish required by the designer. If a medium to heavy exposure of the aggregate is required, a mix omitting the intermediate particle sizes, i.e. the material passing a 10 mm (⅜ in) sieve and retained on a 5 mm (³⁄₁₆ in) sieve, should be used. If only a brush or light exposure of the aggregate is required the mix should be designed in accordance with Clause 3.1. The reason for omitting the intermediate size particles when requiring a medium to heavy exposure of the aggregate is that they prevent the coarse aggregate particles from coming together, with the result that an uneven distribution of the aggregate appears on the face of the concrete. This is not important when only a very light exposure of the aggregate is required. A mix that has been found to give excellent results where a heavy abrasive blasted finish is required is one composed of 1 part cement, 2 parts 3 mm (⅛ in) down fine aggregate and 5 parts 40–20 mm (1½–¾ in) coarse aggregate by weight.*

The strength and hardness of the matrix (cement and fine aggregate) is the important factor, particularly when calling for a heavy exposure of the coarse aggregate. It is obvious that the harder the matrix has become the more difficult it will be to cut it away relative to the aggregate. Tests indicate that for a heavy exposure, 24 hours after placing the concrete is about the limit in normal spring and summer temperatures. For a medium exposure it may be possible to delay the abrasive blasting operation for up to 72 hours. Brush and light abrasive blasted finishes,

Appendix 2

while they can be done at an earlier date, can be carried out up to six months or even longer after casting, depending on the strength of the concrete and the type of coarse aggregate being used. Timing can therefore be critical, particularly when the aggregate is to be heavily exposed and, as no other work can go ahead in the immediate vicinity because of the danger to other workers and because of the dust, it is essential that the work should be most carefully scheduled.

For fuller information on this subject see 'Abrasive blasting of concrete surfaces' (3).

D. BRITISH STANDARDS

BS 2787: 1956	*Glossary of terms for concrete and reinforced concrete.*
BS 3626: 1963	*Recommendations for a system of tolerances and fits for building.*
BS 1881: 1970	*Methods of testing concrete.*
BS 12: 1958	*Portland cement (ordinary and rapid-hardening).*
BS 146: 1958	*Portland–blastfurnace cement.*
BS 882: 1965	*Coarse and fine aggregates from natural sources.*
BS 4449: 1969	*Hot rolled steel bars for the reinforcement of concrete.*
BS 4482: 1969	*Hard drawn mild steel wire for the reinforcement of concrete.*
BS 812: 1967	*Methods for the sampling and testing of mineral aggregates, sands and fillers.*
BS 3148: 1959	*Tests for water for making concrete.*
BS 4461: 1969	*Cold worked steel bars for the reinforcement of concrete.*
BS 4483: 1969	*Steel fabric for the reinforcement of concrete.*
BS 1200: 1955	*Sands for mortar for plain and reinforced brickwork; blockwalling and masonry.*
BS 4466: 1969	*Bending dimensions and scheduling of bars for the reinforcement of concrete.*
BS 1926: 1962	*Ready-mixed concrete.*

BS 2691: 1969   *Steel wire for prestressed concrete.*
BS 2499: 1966   *Hot applied joint sealing compounds for concrete pavements.*
BS 1139: 1964   *Metal scaffolding.*
BS 3589: 1963   *Glossary of general building terms.*
BS 1047: 1952   *Air-cooled blast-furnace slag coarse aggregate for concrete.*
BS 4340: 1968   *Glossary of formwork terms.*
BS 4486: 1969   *Cold worked high tensile alloy steel bars for prestressed concrete.*

REFERENCES

1. SLACK, J. H., and WALKER, M. J. (1967). 'Movement joints in concrete', *Technical Paper*, PCS 29 (London: The Concrete Society).
2. PINK, A. (1967). *Winter concreting* (London: C & CA Publication Cb. 9).
3. WILSON, J. G. *Abrasive blasting of concrete surfaces* (to be published by the Cement and Concrete Association).

Chapter 3 CLAY PRODUCTS

H. W. H. WEST
BSc, FGS, FICeram
*Head of the Heavy Clay Division and
Officer in charge of the Mellor Green Laboratories,
British Ceramic Research Association*

INTRODUCTION

The ceramic external surfaces of buildings may include roofing tiles, tile hanging, floor quarries used on balconies or flat roofs, clay copings used on parapets or free-standing walls, and bricks in brickwork. This chapter is concerned mainly with vertical surfaces and with the properties of bricks, since the phenomena shown by these are also shown in some measure on the pitched or horizontal surfaces. Also, with the exception of frost damage or loss of engobe on roofing tiles, the most important changes which occur during use are seen on brickwork.

Every product has certain properties which convey important advantages in some applications and may be disadvantageous in others. This is as true of materials only slightly modified from their natural state as of those greatly altered by complex manufacturing processes. Ceramic building materials are among the most widely used and certainly the most widely abused products made. Familiarity over 4000 years of use has bred, if not contempt, at least a healthy disregard for any restrictions on use—a brick will always do the job. Everyone knows all about brickwork so there is no need to teach it in schools of architecture—everyone 'knows' that a brick is 9 in $\times 4\frac{1}{2}$ in $\times 3$ in but in fact, of course, it is $8\frac{5}{8}$ in $\times 4\frac{1}{8}$ in $\times 2\frac{5}{8}$ in or now 215 mm $\times$ 102·5 mm $\times$ 65 mm. Seriously though, a product as old in history as a ceramic is naturally expected to be durable, and durable it is, but no one product can be expected to perform equally well in all situations.

A *Class A* engineering brick has very high strength and low water-absorption. It is weather-resistant and chemically resistant. For this reason it is used in situations where it is exposed to extreme conditions from high compressive loading, from weather, or from effluent as in sewer manholes. These desirable properties have concomitant, though minor, disadvantages—relatively high density which makes it heavier to handle, and low porosity and permeability which make it more difficult to lay.

But not all bricks are intended to be made to *Class A* standards. The common brick of this country remains the cheapest manufactured product—at less than £3 per ton delivered, far cheaper than cement even—and although it performs a useful and satisfactory function it cannot reasonably be expected to withstand the extreme conditions for which a product costing several times as much is made.

While the ceramic industries, like others, are continually striving to improve their products so that even extreme conditions of use can be tolerated by an ever wider range of products, and indeed, striving to make new products and designs to meet changing demand, nevertheless there is a standard, BS 3921 (1) to which products must conform but which specifiers must also heed, and Codes of Practice which both designers and constructors should follow. It is common experience that site complaints about the behaviour of brickwork arise because these Standards and Codes have not been followed, and it is appropriate therefore to consider the common faults of brickwork not only in the light of the properties of the product but also in relation to the conditions of use.

FROST

The clay which is processed and fired to make bricks and tiles is formed by the breakdown of hard primary igneous rocks by the physical and chemical agencies of weathering. These same destructive forces—mild seeming enough represented by a gentle summer shower, but remorselessly achieving ultimate breakdown when endlessly repeated over the years—are operating against brickwork, and the most destructive agent of weathering in this country is frost. Winters are invariably wet enough and cold enough for brickwork to become partly saturated with water and then frozen. Ice formation is accompanied by an expansion in volume of 9 per cent and under suitable conditions this expansion may produce stresses which disrupt the bricks, usually causing spalling of a portion of the face.

It will be apparent that in order to take up water a brick must be porous, and many attempts have been made to correlate frost damage in service with the water absorption of the bricks. In fact, of course, there are two measurements of water absorption, that obtained when the brick is soaked in water for 24 hours, and a larger value obtained when it is boiled for 5 hours and weighed after it has been allowed to cool in the water. The difference represents the so-called 'sealed pores', that is, pores which are not directly accessible to water by the normal processes of wetting, but which can be filled by boiling or by soaking under a vacuum. Since these are not filled by soaking in water for 24 hours, they will not be filled by the normal processes of wetting by rain in work. Even if all the open pores are filled with water there will still be some unfilled 'sealed' pore space into which the water can expand on freezing with little or no development of stress. The ratio of the percentage water absorption soaked divided by the percentage water absorption boiled was called the 'Saturation Coefficient', and it was argued that a low value (less than 0·60)—which represented a big difference between the two water absorption values, and hence a large volume of sealed pores—would give good frost resistance. In fact, subsequent work has shown that this is not so, and the 24 hour soak water absorption is no longer a standard requirement in BS 3921 though it may still be used for quality control.

Indeed, it may be said quite generally, that the behaviour of building materials in laboratory freezing tests has, up to now, not shown any correlation with their behaviour in work. It has been suggested that a material of high strength and low water absorption will withstand frost attack and this is true of engineering bricks, since with a limit of 7·0 per cent on the water absorption little water is, in fact, taken up by the brick, and the strength of more than 7000 lbf/in² (49 MN/m²) is sufficient to withstand the disruptive effects of freezing. Nevertheless, bricks of lower strength and high absorption are known to withstand frost conditions satisfactorily and the explanation of this is imperfectly known.

In the laboratories of the British Ceramic Research Association (BCRA) tests have been carried out by three methods:

 a. *Saturation freezing*. The brick is boiled for 5 hours, cooled in water for 24 hours at room temperature, and then placed in a refrigerator for 24 hours.

 b. *Cycling test*. The brick is soaked for 24 hours in water at room temperature then frozen for 24 hours, the test being repeated up to 30 cycles.

Clay Products 109

c. *Zone test.* One end of the piece is kept at −10 °C and the other is held at +25 °C. This produces a plane through the piece at 0 °C and it is at this plane that failure occurs.

This last test was developed in Sweden where the oscillation of the 0 °C isotherm with small changes in temperature was thought to produce the maximum frost damage. Certainly this agrees with the common experience that frost failures are more usually associated with wet winters with temperatures oscillating near freezing, and particularly with freezing rain conditions, than with very severe frosts. Indeed, in Sweden frost damage to roofing tiles is worse in the south where the winters are wet and cold than in the north where the climate is just very cold. Nevertheless, in all our work at the BCRA we have failed to produce consistent frost damage except under completely saturated conditions, and this has been confirmed by other workers.

The same situation is found in work, where damage between damp-proof course (dpc) and eaves is rare and is normally restricted to brickwork which remains saturated, for example, parapets and copings, retaining walls and bricks below dpc. Some of the worst cases occur in boundary walls (Fig. 3.1). Often only common bricks are used, there is no dpc top or bottom and no coping, but only a

Fig.3.1. Frost attack on boundary wall.

capping course of soldiers or brick-on-edge. The selection of materials known to be frost-resistant in similar situations and the provision of correct damp-proofing are important. The standard BS 3921 makes it clear that bricks of ordinary quality are normally durable in the external face of the building, but they are only required to be capable of resisting frost for one winter under the more severe conditions that obtain in a stack on a building site. For conditions of extreme exposure where the structure may become saturated and be frozen, for example, boundary and retaining walls, parapets, sewerage plants, and pavings, bricks of special quality should be specified. At the other extreme, the standard specifies bricks of 'Internal quality' as suitable for internal use only and clearly these should not be used in external work nor left in stack on site during winter conditions.

From the practical point of view it is important to be able to distinguish frost attack clearly from other forms of damage. Confusion arises from lime bursting, which causes pieces to flake off the face, though the white speck of lime which is the culprit can usually be seen; but the chief difficulty is due to handling damage causing chipping, for which frost is invariably blamed. Such handling damage is more usually found at the corners and arisses of the brick and often it can be readily demonstrated because mortar

Fig.3.2. Trowelling damage causing flaking of the face.

is filled into the chip proving it was present when the brick was laid.

A more subtle problem is damage caused by trowelling obliquely across the bed of the brick instead of down the length. This causes a shell-shaped crack to develop on the bed and face and the brick can often be laid without the flake becoming detached (Fig. 3.2). Subsequent weathering, and especially freezing within this crack will cause the piece to spall off and the brick is then said to be suffering from frost damage. It is not the fault of the brick at all, of course, but merely delayed handling damage.

EFFLORESCENCE

Any consideration of durability starts with frost-resistance and the effect of soluble salts, the destructive forms of which cause pitting and spalling which may be mistaken for frost attack. Efflorescence has been recognized since the French Revolution if not as necesarily harmful, at least as a major irritant, detracting from the pride the designer and builder may justifiably feel in the finished structure. The difficulties in controlling it are apparent from the long history of work by many able investigators and although clay products contribute only a proportion, and in modern products often only a minor proportion, of the total soluble salts that cause it, nevertheless, manufacturers of clay building-products are sufficiently alive to their responsibilities in this connection to have made this a continuing part of our research programme over many years. Our investigations have contributed to an understanding of the problem, but extensive research on additives and processing conditions has failed to find one which will render completely insoluble the constituents in all naturally occurring clays.

Efflorescence on external brickwork may be unsightly

but harmless, or may be actually destructive of the facing bricks themselves. On internal walls efflorescence may cause decoration to be delayed or cause damage to decoration already applied.

Efflorescence is the visible effect of crystallization at the surface of salts which have percolated in solution through the brickwork. The amount seen depends upon the quantity and availability of soluble materials and water, and the damage it causes depends upon the chemical nature of the salts. Thus magnesium sulphate which crystallizes just behind the face of the brickwork causes spalling, while other salts, apart from lifting the decoration, are chiefly unsightly.

The salts may be derived from the walling units themselves, from the mortar and plaster, or from contamination from some source other than the wall.

Most clay building-materials contain a small percentage of salts soluble in water, commonly the sulphates of calcium, magnesium, sodium, and potassium. In a few materials ferrous sulphate may be present. Various investigators in different parts of the world have identified also molybdenum, nickel, chromium, and vanadium compounds as soluble salts in isolated samples, but these occur too rarely to be considered here,

The soluble sulphates may have been present in the clay, occurring, for example, as gypsum (calcium sulphate), or may be formed during the firing, either by the oxidation of pyrites or by reaction between sulphurous gases (formed by the oxidation of the sulphur in the fuel) and calcium and magnesium carbonates present in the clay. Some sulphates are decomposed at high temperatures and well-fired bricks therefore tend to contain a lower total content of soluble salts than under-fired bricks made from the same clay.

The presence of soluble salts in a brick does not necessarily mean that efflorescence will ensue. This depends, among other things, upon the solubilities of the salts concerned. For example, calcium sulphate, which makes up the greater part of the soluble salts in many materials, has a relatively low solubility. This apart, however, some bricks with a high content of soluble salts show no efflorescence.

BS 3921 specifies that no brick of any quality should develop efflorescence worse than 'moderate' which means that when a sample of ten is tested in such a way that water is allowed to evaporate from one face only, up to 50 per cent of the area of that face may be covered with a deposit of salts, but it should be unaccompanied by powdering or flaking of the surface.

No limitations are placed on the content of soluble salts, however, except in the case of special quality bricks where the contents by weight per cent of the soluble radicals shall not exceed the following:

Acid soluble sulphate	0·30
Calcium	0·10
Magnesium	0·03
Potassium	0·03
Sodium	0·03

Clearly where it is imperative that the soluble salt content of the bricks is at minimum then 'Special quality' should be ordered.

Even if facing bricks of low soluble salt content are used, however, salts in backing bricks, or breeze or concrete blocks, may be taken into solution and migrate through the wall to appear as efflorescence at the drying surface. This and other sources of efflorescence are shown in Fig. 3.3.

Portland cement and hydraulic line mortars may provide soluble salts, usually the sulphates and carbonates of sodium and potassium. The use of magnesian lime in mortars may provide a source of magnesium which could lead to the production of destructive magnesium sulphate efflorescence in bricks that would otherwise be satisfactory.

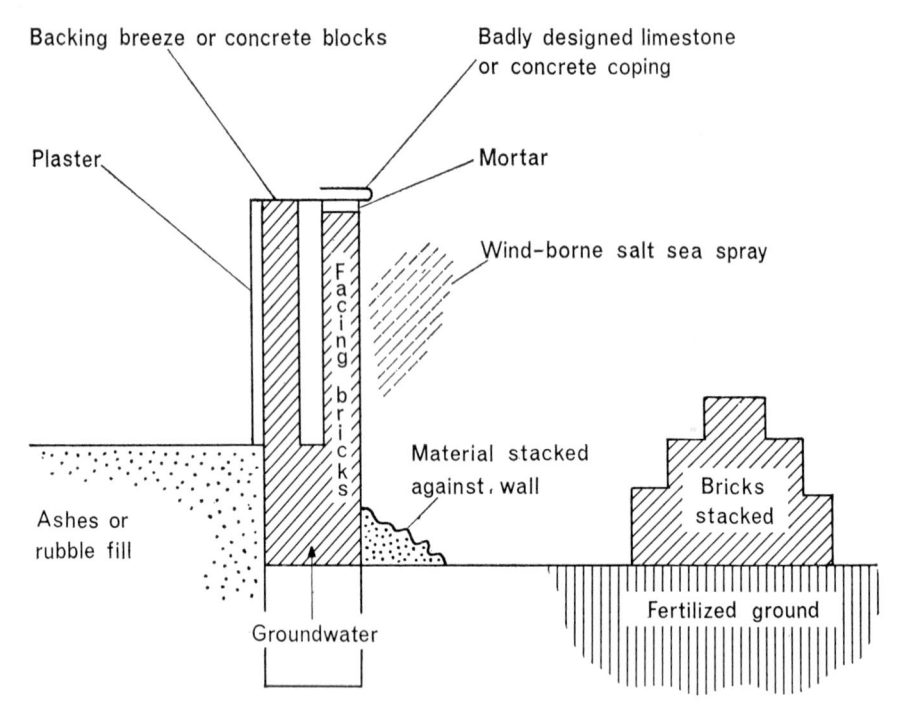

Fig. 3.3. Sources of soluble salts.

[Diagram labels: Backing breeze or concrete blocks; Badly designed limestone or concrete coping; Plaster; Mortar; Wind-borne salt sea spray; Facing bricks; Material stacked against wall; Bricks stacked; Ashes or rubble fill; Groundwater; Fertilized ground]

This may result, for example, from the reaction of the magnesium carbonate with rainwater containing dissolved sulphur trioxide from industrial gases.

In internal work, soluble salts are present in the plaster and may give rise to efflorescence on their own account. Indeed, BS 1191: 1967, *Gypsum building plasters* (2), places a limit of 0·2 per cent on the sum of the content of soluble sodium and magnesium expressed as Na_2O and MgO.

Contamination from outside may occur from groundwater, from substances stacked against the wall, or even from spray borne inland from the sea. Materials may pick up salts before use if stacked on ground contaminated by, for example, fertilizers. The presence of nitrates or chlorides in the efflorescence is usually an indication of this sort of external contamination. In one case of recurring potassium nitrate efflorescence in floor quarries, it was found that the tiles were laid on concrete which had been placed on prepared ground containing material from a demolished larder in which pigs had been salted with saltpetre.

Soil, and particularly ashes and rubble filling, behind retaining walls may contain high proportions of soluble salts which appear on the brickwork and perhaps destroy it.

Rainwater will take calcium carbonate or calcium sulphate into solution from limestone or concrete copings, etc., and if this solution does not fall clear of the building it may be taken up by the brickwork to give rise to efflorescence on subsequent drying out.

The place at which efflorescence occurs is no certain guide to the provenance of the salts. The solutions may have migrated considerable distances through the complex of the wall, and their appearance on a particular material merely indicates that it provides a convenient drying surface. The use of a very dense impermeable mortar pointing in conjunction with a more permeable brick will allow water to flow into the wall mainly through the brick and dry out the same way, so that efflorescence will appear on the brick although the salts may have arisen from the brick or mortar or both.

For efflorescence to occur in brickwork, salts must be present, water must be available to take them into solution, and a drying surface must exist at which evaporation can proceed to deposit the crystals at the surface. If no water were available, efflorescence could not occur, but water is, of course, provided by the mortar during the process of laying bricks, by rain before tiling, and in external brickwork water is continually provided by the weather. Although efflorescence can occur on the plaster of internal walls, if this is removed, once the brickwork and plaster have dried out no further efflorescence will occur.

A special case of damage due to soluble material is

brown staining of plaster. Black-hearted bricks contain iron in the reduced ferrous state and this is soluble in the slightly acid rainwater. During the course of building, therefore, the ferrous iron can be taken into solution and as the building dries out the solutions migrate to the drying surface. If a lime sand backing coat has been used, no difficulty arises, because when the acid solution is rendered alkaline by contact with the lime, the ferrous iron precipitates as a brown ferric iron layer at the interface between the brick wall and the backing coat. When, however, lime-free plasters are used, there is no alkaline barrier and the solution dries out at the outside surface of the plaster giving a brown stain which spoils the decoration.

When this point has been reached, the stain must be hidden by paper or dark paint, though it is important to ensure that the wall has dried out completely before re-decorating. If it is known that the bricks are black-hearted, and the work is likely to be left uncovered, then a barrier should be provided either by specifying a lime backing coat or alternatively by merely lime-washing the brickwork before applying the plaster.

Brown staining may also be found on the mortar joints of facing brickwork built of bricks containing soluble iron. The only certain solution is to keep the brickwork dry during erection and for as long as practicable afterwards, since the staining only occurs when the solutions meet fresh mortar, and after 3–7 days discoloration is less likely.

The weather during erection greatly influences the formation of efflorescence. Buildings erected during the winter often show the phenomenon when the brickwork dries out in spring. Covering the work and the stacks of material during erection will minimize efflorescence, but once the building has shown it the only treatment is to brush the deposit off. Washing has the effect of dissolving the salts and the solution is absorbed as it runs down the brickwork and the efflorescence appears in another place on subsequent drying.

Fig.3.4. Efflorescence on gable end.

If, after taking all precautions during construction, efflorescence recurs in dry periods following rain, this usually means that excessive water is entering the wall through faulty detailing, for example, parapet walls with no damp-course, or with one that has been wrongly placed. Fig. 3.4 shows massive efflorescence on a gable end due to bad detailing of the roof.

DIMENSIONAL CHANGES IN BRICKS AND BUILDINGS

All building materials expand and contract and these changes in dimensions are not in themselves harmful. It is the combination of different movements due to the conjunction of dissimilar materials in a building which may give rise to difficulty. In addition, of course, the building as a whole is moving due to changes in the live loads, to wind forces, to settlement or even subsidence in the foundations. The possibility of these movements occurring is well known, the important thing is to ensure that the building is designed to accommodate them without damage.

While many examples of old brickwork may be seen in which no distress due to movement is evident, particularly long estate boundary walls, it should be recognized that these walls were built in lime sand mortar able to accommodate quite large movements without distress. Higher strength cement mortars, however, not only provide less flexibility in the joints, but also themselves shrink on setting and thus may lead to cracking. Indeed, drying shrinkage is to be expected of all cementitious products. In all modern building, therefore, it is imperative to design and install proper movement joints when the properties of the material or the dimensions of the building are such that damage may result from movements which can be predicted from prior knowledge.

Thermal movement

The coefficient of thermal expansion of building materials varies with the particular sample, but approximate ranges are given in Table 3.1.

Table 3.1. *Coefficient of linear thermal expansion of building materials in the range 0–212 °F (−18 to 100 °C).*

Material	Coefficient of linear expansion $\times 10^6$ per °F	per °C
Bricks, brickwork	2·8–4·0	5·0–7·2
Lime mortar	4·1–5·1	7·4–9·2
Cement mortar and concrete (depending on mix proportions and type of aggregate)	3·2–8·1	5·8–14·6
Steel	6–7	10·8–12·6

(After BRS (3).)

Seasonal variations of temperature may be over 50 °F (28 °C) in this country, but from the structural point of view diurnal variations are more important. For example,

Clay Products

a change of 20 °F (11 °C), which in summer could occur quite rapidly, would give rise to a dimensional change of 0·1 in (2·54 mm) in a brick wall 100 ft (30·5 m) long if unrestrained. CP 121 (4) recommends that the effect of thermal stresses should be considered in brickwork over 100 ft (30·5 m) long. Indeed, expansion joints should be provided every 40 ft (12·2 m) approximately. In practice structures are restrained so that the expansion develops stresses which may be relieved in severe cases by cracking of the brickwork or of the restraining element.

Thermal expansion is theoretically reversible, and is so in the case of an individual brick. In brickwork, however, the wall often expands horizontally by sliding on the dpc and will then only return to its original position if it remains uncracked. Vertical thermal movements are generally reversible and should be taken as being 1½ times the values given in Table 3.1.

Thermal expansion effects are usually designed for in long walls, but it is important to recognize that in most, if not all, cases attributed to moisture expansion, an element of thermal expansion is present. Great care is invariably taken in designing steel structures to make adequate allowance for thermal movement. Note that in Table 3.1 the coefficients of expansion of brickwork and concrete are of the same order as that of steel.

Moisture movement

All porous building materials exposed to the weather take up water. Expansion of materials is associated with wetting and shrinkage with drying. Cement products show, in general, an initial contraction during setting, followed by expansion and further contraction on subsequent wetting and drying, but some proportion of the initial shrinkage is irreversible. Clay products also show expansion on weting and contraction on drying, the expansion being quite rapid initially and continuing at a much lower rate over very long periods. The expansion is not wholly reversible on drying. The net effect in conditions of varying humidity, once the initial rapid expansion is completed, is a very gradual continuing expansion.

The magnitude of these movements varies with the material. Measurements are made on unrestrained pieces, but in the wall much of this movement is restrained. Nevertheless, some movement remains and in long walls must be accommodated by suitable expansion joints and attention to detail—for example, short returns should be avoided to prevent cracking adjacent to the quoin. The BRS (5) have found such cracking mainly when the length of the return is $13\frac{1}{2}$ in (343 mm) or less, none having been found when the length is above 2 ft 3 in (686 mm).

The magnitude of the moisture expansion of individual bricks is difficult to replicate, and for this reason no method of measurement is given in BS 3921. Bricks straight from the kiln show an initial rapid moisture expansion which, if measured at this stage, may be up to 0·15 per cent or 0·2 per cent linear, but under normal conditions of handling on works and site much of this has taken place before the bricks get into work, and the residual expansion is minute by comparison.

The determination of the moisture movements of bricks and briquettes is challenging enough, but in brickwork it is complicated by interaction between the bricks and mortar. Indeed, this whole problem of the magnitude of expansion of brickwork and of its importance as a defect needs careful examination. It is true that short returns in long walls without expansion joints may crack—they might in some circumstances crack by thermal movements alone—but the difficulty seems to be to provide a reasonable opportunity for moisture expansion to take effect. The intial, very rapid and large, expansion takes place immediately ex-kiln, and there cannot be many occasions when it has not taken place before the bricks are laid. Even if the bricks are very new when they reach the scaffold, the effect of the water from the mortar

Clay Products

can only be to cause rapid expansion. This expansion must go on as water continues to be withdrawn from the mortar, and the latter is surely not sufficiently rigid to restrain the expansion of the bricks. It would seem then that the early expansion, certainly in the first 24 hours must take place without harm to the brickwork. Indeed, movement may be accommodated for some time by the mortar, probably until seven days in any case, and perhaps for longer with weak lime mortars.

Certain recommendations can be made to minimize problems due to moisture expansion. Kiln-fresh bricks should not be used without wetting, excessively rigid mortars should be avoided, and movement joints should be provided to cope with both moisture and thermal movements in all building products.

Shrinkage of reinforced concrete frames

There have been in recent years in the United Kingdom a number of cases of severe cracking and spalling of the facing brickwork cladding of multistorey reinforced-concrete-frame buildings. Such cases are often quoted as examples of differential movement—specifically moisture expansion of the bricks has been blamed—but it is difficult to ascribe forces to the bricks large enough to account for the effects seen, particularly under such restraint. It is an interesting observation that buildings of this kind usually show vertical rather than horizontal movement, which points to a shortening of the reinforced-concrete-frame.

In this form of construction it is usual to overhang the outer leaf one-third of its thickness and to cover the outer edge of the floor slab with brick slips so that an over-all appearance of brick is maintained. This is a practice stemming from old LCC regulations and one used widely for the substantial brick-clad steel-frame structures of the inter-war years. In these, the outer walls were nearly always $13\frac{1}{2}$ in (343 mm) thick (nominal) and, therefore, could overhang $4\frac{1}{2}$ in (114 mm) (nominal). Thus any part

122 PERFORMANCE OF BUILDING MATERIALS

Labels (left diagram):
- Back line of RC 10in x 10in column
- Floor
- dpc
- 1in slip
- T/3 Where T = Outer leaf thickness
- Clay block inner leaf

One form of modern construction

Labels (right diagram):
- T
- Stanchion
- Floor
- T/3

circa 1930 Steel frame construction (old LCC regulations)

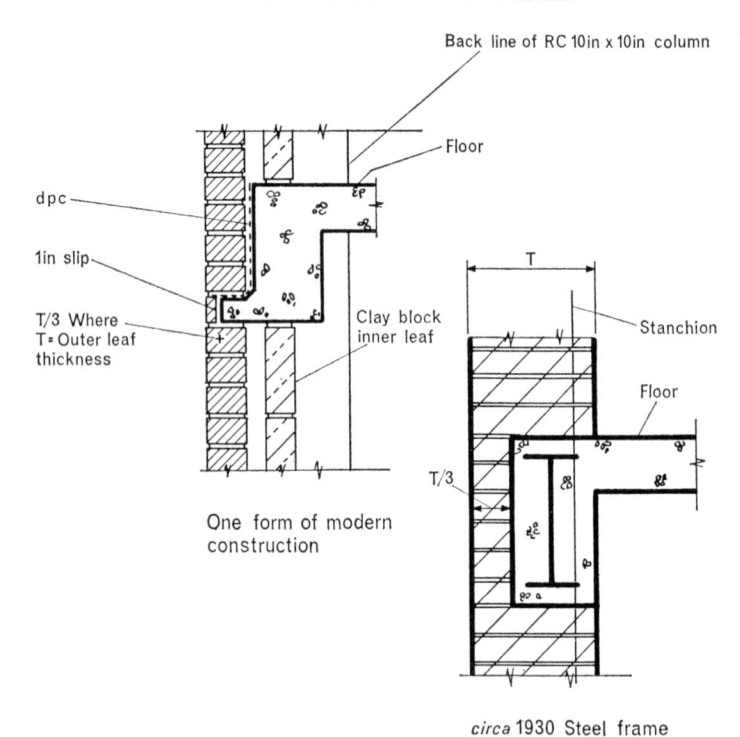

Fig.3.5. Comparison between inter-war and present arrangements of brick cladding on frame structures.

Fig. 3.6. Spalling of brick slips covering the toe of a concrete floor.

Clay Products

of the cladding in front of a spandrel beam or floor edge was of substantial thickness and no trouble of the kind described below occurred. The reasons for the difference in behaviour are that on the one hand there is a non-shrinkable, low-creep, steel frame to which a substantial cladding is firmly fixed, and on the other there is a reinforced-concrete frame of high-shrinkage, high-creep characteristics to which a slender cladding is less substantially attached. Economy has changed the overhang from one-third of a substantial thickness to one-third of a very slender skin (Fig. 3.5)(6).

In early reinforced-concrete-frame buildings, and indeed in some later ones built when adequate knowledge was available, no movement joints were fitted between the top of the brick cladding and the concrete floor level above. Thermal expansion of the brick skin must in any event be reckoned with and, although moisture movement of all porous building materials occurs, the expansion of the ceramics is small compared with the drying shrinkage of the reinforced-concrete frame. Apparently, little is known about the mechanism of linear (vertical) shrinkage of concrete columns when these are reinforced, but there is ample evidence that it takes place to a marked degree. Shortening recorded in the USA has been of the order of 1 in per 100 ft (25 mm per 30 m) in heated structures. The effect of these differential movements is that an eccentric load is unintentionally placed upon the outer leaf, and is taken on the inner 3 in (76 mm) or so.

Failures of the brick cladding have been seen. Often the brick slip course covering the toe of the concrete floor will spall off, there may be spalling of the face of one or two courses of bricks, and vertical cracks may occur. All these are symptoms of a high compressive loading on the brickwork due mainly to the shrinkage of the concrete frame (Figure 3.6).

That this defect has only become apparent in the last decade seems to be due to a combination of better internal space heating and more slender design. For perhaps the first time since they were first used, reinforced-concrete elements are really drying out to the extent that the aggregates themselves are shrinking, and it is now known that some aggregates shrink markedly when the concrete fully dries.

Differential thermal and moisture movement of the cladding cannot be entirely eliminated, but it is significant that similar spalling has taken place in stone-clad, reinforced-concrete structures, yet moisture expansion of stone is virtually unknown.

MORTAR DECAY DUE TO THE FORMATION OF CALCIUM SULPHO-ALUMINATE

Portland cement containing tricalcium aluminate reacts with sulphates in solution to form calcium sulpho-aluminate which causes expansion and loss of strength in the mortar. Persistently damp conditions are necessary for this to occur, so mortar between dpc and eaves does not often suffer from this defect except in regions exposed to driving rain. Dense renderings, however, may crack and allow water to penetrate behind them, producing conditions in which decay can occur.

On long stretches of brickwork the expansion may be seen as oversailing of the dpc since sulphate expansion is usually less below dpc possibly due to restraint by the foundations. At this time care is necessary to distinguish between sulphate attack and thermal and moisture expansion. As the attack continues the mortar joints may crack. Later on the surface of the joint spalls off and the mortar becomes soft and crumbly and is readily removed from the joint (Fig. 3.7). Spalling of the faces of bricks may occur either due to compression effects or to the

Fig.3.7.
Deterioration of mortar due to the formation of calcium sulpho-aluminate.

crystallization of salts behind the face. In rendered brickwork, the rendering shows horizontal cracks due to expansion of the brickwork behind and eventually extensive areas of rendering may fall off.

Since the cause of the defect is the presence of tricalcium aluminate in Portland cement, it can be entirely avoided by the use of sulphate-resistant cement in which the quantity of tricalcium aluminate, which is the least stable compound, is strictly limited.

Tricalcium aluminate, soluble sulphates, and water must all be present, and it would be logical to restrict the occurrence of the defect by specifying sulphate-resistant cement, which can be provided, rather than by restricting the soluble salt content of the bricks which is not the critical factor in situations where migrating solutions may bring contamination from elsewhere. Nevertheless, CP 121 recommends that for parapets, copings, external free-standing walls, and retaining walls, the soluble sulphate content of the brick should be less than 1 per cent, and less than 0·5 per cent in very wet conditions, unless there is evidence that the particular brick will be satisfactory. In fact, bricks used in these situations should be of 'special quality' as defined in BS 3921, and this places a more stringent limit on the soluble sulphate content.

The worst cases of sulphate attack, however, occur in garden walls with no dpc and here the original soluble salt content of the bricks is much less important than the salts picked up from the soil. CP 121 also recommends that detailing should be arranged to avoid brickwork becoming and remaining wet. Dense renderings are to be avoided, and details at eaves, verges, sills, parapets, and copings should be designed to throw water clear of the brickwork.

DESIGN AND CONSTRUCTION

So far we have considered some of the properties of building materials which influence their resistance to weathering. The durability of walling is not solely determined by the quality of the external bricks, but depends upon the interaction of the whole brickwork complex, including mortar, rendering if any, and backing material. Good design and detailing is all important in determining the conditions of exposure to which the brickwork is subjected, and thus much of the credit, or blame, for the eventual behaviour of the building must rest with the designer and constructor.

Perhaps, because of familiarity, we think of buildings as simple structures. In fact, the operation of building is complex if only because of the complex nature of the construction team. The architect sees his design as the critical solution to the demands of the client within the economic limits set. His detail drawings are intended to be the practical solutions to the problems of turning his design ideas into a finished structure. The builder or contractor is concerned to erect the building for the minimum cost within the terms of his contract. It has been well said that there are accordingly three buildings:

> The one the architect thinks he designed.
> The one the architect actually did design.
> The one which the builder finally erects.

Clay Products

The client *commissions* the architect who *instructs* the builder who *employs* sub-contractors who *buy* materials. In any failure in a building blame starts at the bottom of this list and rarely reaches the top where in at least some cases it should really lie. In consequence, therefore, any failure of brickwork is regarded first as a failure of bricks, and a complaint is made to the brick manufacturer. Often the fault lies in a failure to design correctly, or is due to inadequate or even frankly bad workmanship. It is appropriate to consider the responsibilities of the various parties in that process which has been described as starting with a sketch on the back of an old envelope and ending with a cathedral!

The specifications for materials are given in the appropriate British Standards and the one with which we are most concerned is BS 3921 : 1969, *Bricks and blocks of fired brickearth, clay or shale*. The statutory requirements of design and construction are, of course, laid down in the Building Regulations and various by-laws, but the recommendations of good practice which should be followed by both designer and builder are incorporated into the Codes of Practice for Buildings. The one which chiefly concerns us now is CP 121 : 101 : 1951, *Brickwork*, for which a revision is in draft.

Specification for bricks

The BS Specification lays down the requirements of quality which the manufacturer must meet. In summary he is responsible for supplying bricks:

    a. To size;

    b. To strength;

    c. To a standard of finish;

    d. To pass the efflorescence test as moderate;

and also where applicable:

    e. To a defined frost resistance;

    f. To a defined soluble salt content.

These are the limits of the manufacturer's liability, but if test data are given to the user in support of compliance with the Specification then such tests must have been carried out in accordance with Part 3 of the Standard on a sample obtained by the methods described.

The designer's responsibilities

CP 121: 101: 1951 places responsibility for the choice of materials on the designer noting especially that bricks for external facing work 'should ... be of a quality suitable for the conditions prevailing at the most exposed positions of the particular job' (Clause 204(a)).

A Code of Practice is not mandatory; indeed Codes carry a note, 'Compliance with it does not confer immunity from relevant legal requirements, including by-laws', but it does represent 'a standard of good practice and therefore takes the form of recommendations', It follows that if the design is such that the durability of bricks of ordinary quality is adversely affected, then the responsibility is with the designer for failing to carry out the recommendation above and not with the brickmaker who supplied goods to specification.

Brickwork between dpc and eaves is not, however, a major source of complaint, and the Code is quite specific about those other areas of brickwork from which trouble often ensues, and points out that parapets, copings, external free-standing walls and retaining walls are exposed to more severe conditions and thus bricks of superior quality are necessary at these positions.

The designer's first responsibility is to make the structure stable and safe. This is done either by designing on the basis of specified permissible stresses when reference should be made to CP 3, Chapter V (7) and to CP 111 (8) or, when the thicknesses of walls are to be determined on the basis of length and height only, to the relevant building by-laws of the local authority concerned.

In this country the designer's responsibility to exclude

rain is of prime importance, and the degree of exposure to which the building as a whole or parts of it (notably parapets and chimneys) is subjected must be determined. The exposure conditions are defined as three Groups in the Code as 'Sheltered', 'Moderate', or 'Severe' (4). While '... unrendered solid brickwork should not be relied upon to give satisfactory protection against rain except under sheltered or moderate conditions ... a soundly constructed hollow wall provides an absolute barrier to rain penetration ...'.

Brickwork does not have to be waterproof, but it should be weather-proof. When well-vitrified bricks are used rain is shed and runs down the face of the building, but when porous bricks are used the water is soaked up and evaporates again when the rain stops and drying winds or sun appear. These two importantly different ways in which brickwork excludes rain have been well dubbed the 'raincoat effect' and the 'overcoat effect' respectively.

Only in very rare cases of highly permeable material does rain ever pass through the bricks in a wall. The most usual path of entry is through incompletely filled perpend joints or through cracks in the bed joints. With bad workmanship of this kind, in conditions of severe exposure water can be sprayed across the cavity to wet the inner leaf.

The prevention of damp in a building is also dependent on the provision of proper dpc both horizontal and vertical. Clause 313 not only lays down where dpc are to be provided, but also gives illustrations of the correct positioning of dpc in various situations.

The designer's responsibility for the provision of movement joints is perhaps stated somewhat less than adequately as being '... desirable to consider the effect ... of temperature' in walls over 100 ft (30·5 m) long. In fact, it is now recognized that differential movement must also be considered, the shrinkage of concrete in, for example, framed buildings as well as possible moisture expansion of clay bricks, so that movement joints ⅜ in (9·5 mm) wide containing easily compressible material should be incorporated every 40 ft (12·2 m).

In summary the designer is responsible for:

a. The correct choice of materials;

b. The strength and stability of the building;

c. Exclusion of rain;

d. Prevention of moving damp by the correct detailing of dpc;

e. Provision of movement joints.

The Contractor's responsibilities Section 5 of CP 121: 101: 1951, *Work on site*, provides a guide to some of the operations of building and gives advice which, if followed, would minimize complaints. Sub-contract bricklaying tends to be of as low a standard as site supervision will allow. Building materials manufacturers may not be entitled to demand a better standard of workmanship, but they are entitled to demand that their product is not blamed for faults which are due to bad workmanship.

The Code clearly lays down that new brickwork must be protected against rain, snow, and frost. Complaints are often made that if perforated bricks are cut out of a wall water runs out 'like cracking a coconut' and this is said to 'prove' that the brickwork is permeable and the perforations are filling with rainwater. Extensive investigations on bricks of numerous perforation patterns and water absorption have shown that the quality of the workmanship and the performance of the mortar are the most important determinants of rain penetration through brickwork. Tests of rain penetration in which the workmanship was deliberately faulty have confirmed the views of other workers that this is crucial to good rain-resistance, and the presence or pattern of the perforations is of no significance. Although it might well be thought possible for water entering through cracks in the mortar joints to collect in the perforations this does not seem to happen, and water found in perforated brick walls is indicative of the work having been left uncovered during the course of construction.

The most serious cause of complaints about rain penetration is, however, undoubtedly due to bad construction of cavity walls. The Code is absolutely clear on this point and no excuse should be accepted for failure to comply.

It will be apparent that since a properly constructed cavity wall is an absolute barrier to rain penetration, any complaint directed to the brick manufacturer about damp on the inner leaf must be frivolous. The causes may be brickwork saturated during building and not fully dried out before occupation, bad workmanship shown by open mortar joints which allow water to spray across the cavity, or dirty cavities which allow water to bridge from the outer to the inner leaf or to rise from the foundations.

Isolated patches of damp randomly distributed over the wall are indicative of a bridged cavity, usually dirty wall ties. If the damp patches are just above ground floor level the likelihood is that mortar droppings have built up above the dpc to provide a bridge and route for rising damp. The dpc itself may also be suspect. Damp patches over openings or on reveals are indicative of faulty or absent dpc.

Chimney stacks are a fruitful source of complaints of water penetration. Above roof level they may be exposed to severe conditions, and tall stacks give more difficulty due to the greater catchment area of the brickwork. Water penetrates the exposed brickwork and runs down the inside of the stack between the brickwork and flue liner giving rise to damp and stained areas on the chimney breast.

CP 131: 101: 1951 (9); *Flues for domestic appliances*

burning solid fuel, gives complete guidance on the dpc to prevent this. Rarely are the provisions of this Code followed. Indeed, the wrong location and installation of dpc are perhaps the most fruitful source of penetration of water into a building.

Finally, the Code requires that materials and workmanship be inspected. If we could always be sure that this clause was carried out, then we could also be sure that the compliment paid to brickwork in the last clause of the Code would be justified. 'Brickwork built in accordance with the recommendations of this Code should, in ordinary circumstances, need little maintenance.'

CONCLUSIONS

It is on just this property of freedom from maintenance that the long continued success of clay building-products depends. Of course, one can quote the aesthetic appeal of good brickwork, the architectural appeal of the textures it makes possible, the charm of old brickwork and clay roofing tiles which weather gracefully under any conditions, but in the end the brutal economic facts of life are that brickwork is cheap to lay and it costs nothing to maintain, except perhaps for occasional repointing.

This chapter has tried to set down those properties which are important to the behaviour in work of clay products: frost-resistance, freedom from efflorescence, the dimensional changes which are bound to occur, and the possibility of failure of the mortar due to the formation of calcium sulphoaluminate. While these properties must be taken into account, the responsibilities of supplier, designer, and user are clear. If they accept these responsibilities and all supervise the workmanship for which they are responsible, then there need be no cause to fear the performance of the external surfaces of ceramic-clad buildings.

REFERENCES

1. BRITISH STANDARDS INSTITUTION (1969). *Bricks and blocks of fired brickearth, clay or shale*, BS 3921: Part 2.
2. —— (1967). *Gypsum building plasters*, BS 1191.
3. BUILDING RESEARCH STATION (1959). *Principles of modern building*, vol. 1, 3rd edn (London: HMSO).
4. BRITISH STANDARDS INSTITUTION (1951). *Brickwork*, CP 121: 101 (revision in publication).
5. BUILDING RESEARCH STATION (1958). *BRS Digest*, no. 114 (First Series) (London: HMSO).
6. FOSTER, D. Private communication.
7. BRITISH STANDARDS INSTITUTION (1967). *Dead and imposed loads*, CP 3: Part 1, chapter V.
8. —— (1964). *Structural recommendations for loadbearing walls*, CP 111.
9. —— (1951). *Flues for domestic appliances burning solid fuel*, CP 131: 101.

Chapter 4 TIMBER

GAVIN S. HALL

BSc, MF, DFor, AIWSc
Manager, Wood Technology Department,
Timber Research and Development Association,
Hughenden Valley, High Wycombe, Bucks.

INTRODUCTION

In many countries of the world, timber is a traditional material for the external cladding of buildings. This is in marked contrast to the situation in the UK where it has been widely used only in recent years under Scandinavian and North American influence. However, much timber has traditionally been used for trim and exterior joinery on buildings built of brick or stone. The post-war period has seen a departure from ingrained building designs and practices, and this movement has involved the use of a wide variety of alternative materials for these exterior surfaces. Timber, finished either naturally or with a paint system, has come to be used in a variety of ways including cladding, infill panels, and in curtain walling systems. Its performance in these applications is the subject of this chapter.

Timber is the generic term for planks, baulks, scantlings, etc., sawn from logs. There are other ways in which logs are converted into building materials. Most important of these is plywood. The special characteristics of this product, together with those of particleboard and fibreboard, will be considered later. Materials produced by a process other than sawing, for example, slicing or chipping, are usually referred to as 'wood-based'. The processing is essentially mechanical and many of the basic properties of the raw material are retained. This raw material is 'wood' and by way of introduction here is a brief account of its chemical and physical constitution and properties.

THE CHEMICAL AND PHYSICAL NATURE OF WOOD

Chemically, wood is a compound polymer with three main constituents. On average, 50 per cent by weight of wood consists of cellulose, a linear polymer responsible for timber's strength, elasticity, and toughness. The long chain-like molecules of cellulose are grouped into bundles which run helically to form hollow, needle-shaped cells or fibres, 1–3 mm in length. Associated with these cellulose bundles are a variety of similar polymeric materials collectively termed hemicelluloses whilst binding the bundles together is a branched polymer cementing material known as lignin. Lignin is also responsible for binding the individual fibres together to make up wood. The manufacturer of paper involves the removal of most of this binding agent and the liberation of individual wood fibres which are then felted together to produce the sheet material. In the raw state, wood contains about 25 per cent lignin (1).

The basic composition of wood varies slightly. For instance, *softwoods* (timber from coniferous trees which is not necessarily soft) contain more lignin and less hemicellulose than most *hardwoods* (broad-leaved trees yielding timber which may be very hard (greenheart), or very soft (balsa)). This variation has little influence on the properties of timber plywood, chipboard, etc., and only assumes significance when chemical processing is involved.

The microscopic structure of wood is much more important than its chemical composition in determining the properties of a timber. This structure is related to the functions of wood in a tree, namely mechanical support and water conduction from the roots to the leaves. In softwoods, the microscopic structure is relatively simple because the same type of cell performs both functions. In cross-section, a piece of softwood looks like a series of boxes cemented together. Viewed sideways, these cells look rather like minute toothpicks, being 20–50 times longer than their width. A similar type of cell is found in hardwoods but here the fibre is shorter and the cell wall thicker. Other large-diameter cells, termed vessels, serve the water-conducting function in the living tree. The microscopic organization of the cell wall is that which determines the properties of wood but, at the microscopic level, the main concept needed for the understanding of the behaviour of wood exposed to the weather is that of the cellular construction.

A few macroscopic features also need to be pointed out before embarking on a discussion of wood's reaction to exposure. Perhaps the first requirement is to define the terms transverse, radial, and tangential. To understand these it is necessary to think back to the tree. Diagrammatically, a tree can be considered as a series of cones superimposed as shown in Fig. 4.1, with a new cone being added to the outside with each season's growth. In transverse section the diagrammatic log looks like a series of concentric circles. The tangential plane is one at a tangent to these circles while the radial plane is at right-angles to it and passes through the centre. Planks sawn tangentially, or largely so, are termed flat-sawn, while radially cut pieces are often termed quarter-sawn.

Fig.4.1. Diagrammatic sections through a tree showing growth-rings and radial and tangential axes.

Longitudinal section

Cross-section

Radial axis

Tangential axis

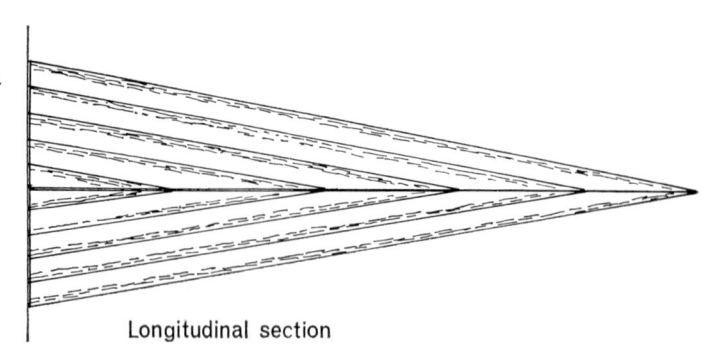

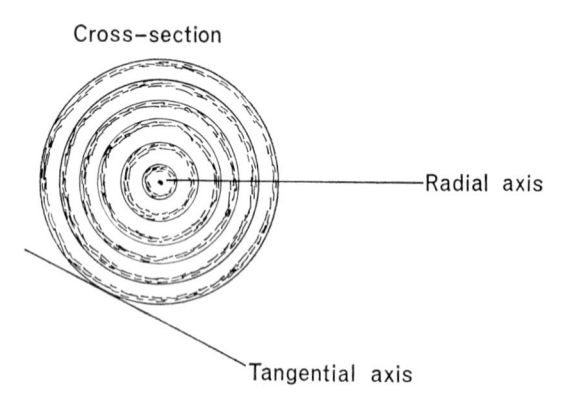

Fig.4.2. Flat-sawn and quarter-sawn boards.

Flat or plain-sawn

Edge or quarter-sawn

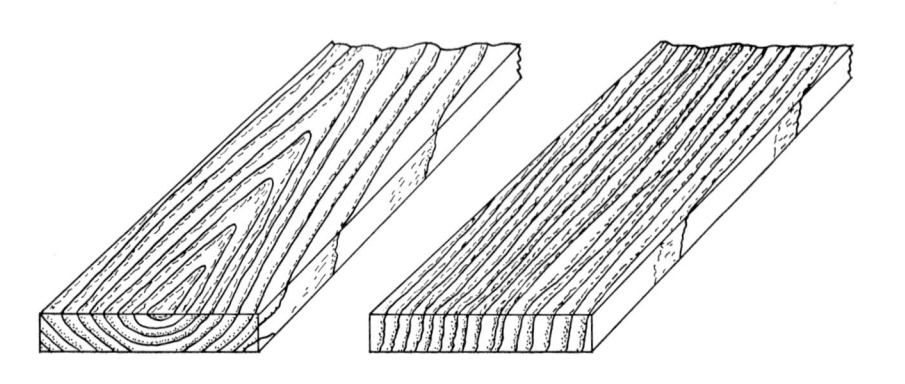

Timber

In conifers, the thickness of the cell walls, and in hardwoods, the proportion of the different types of cells varies during a seasons growth (in many tree species) and the concentric circles of Fig. 4.1 are visible as growth-rings. Whatever the cause, the latewood is usually denser and harder than the earlywood or springwood formed at the beginning of the growing season. This contrast gives rise to the typical appearance of quarter-sawn and flat-sawn timber (Fig. 4.2). With tropical hardwood species, with no such distinction between early- and latewood, it is difficult to determine the orientation from superficial observation.

The wood of different timber species differs widely in physical and mechanical properties, largely as a result of variations in density. Aside from this, much of the individuality of the wood of a particular species is due to the extractives contained in it. These extractives are loosely bound chemicals which can be removed with inert solvents such as water, alcohol, benzene, etc.; hence the term 'extractive'. Chemically, extractives are complex and each species has its chemical 'fingerprint' which has occasionally been used to separate woods with otherwise indistinguishable structures. These chemicals are responsible for most of the colour of timber although the lignin component imparts a slight straw-colour. Also they make a timber resistant to decay and insect attack to varying degrees depending on their type and quantity. They also give a species of wood its characteristic smell and influence its water-resistance and shrinking and swelling properties. Their effect is out of all proportion to their amount since most timbers contain only 1–2 per cent extractives by weight.

WOOD EXPOSED TO THE WEATHER

Of the weathering influences, to which wood exposed outside is subjected, only water and light are significant.

The thermal expansion of wood is small and not an important cause of degradation whilst wood has a deservedly good reputation for resistance to chemical corrosion and is largely unaffected by atmospheric pollutants. The exception is a change in appearance due to dirt deposition. Wood does not 'age' on exposure to air alone.

Since wood is largely cellulosic and cellulose is a hygroscopic material, it follows that water must be part of its constitution. The affinity of cellulose for water is extremely high and continued heating above 100 °C is necessary to remove it completely. Under normal conditions, wood contains some 12–18 per cent by weight of water. This moisture content is mainly controlled by the relative humidity of the surrounding air and is largely independent of temperature.

Colour changes If a freshly surfaced piece of timber is exposed to the weather, the first change is in its colour. Most dark-coloured timbers will begin to bleach whereas pale timbers often darken and redden slightly before beginning to lose their colour. Certain species such as iroko and afrormosia show a very marked colour change on exposure to light. In all cases, photo-oxidation of the extractive compounds and lignin is responsible and, although the changes are much more rapid on exterior exposure they also occur in wood used for interiors. Here colour stability and particularly colour uniformity is important and a commercial process for pretreating timber to reduce the change on exposure to light has been developed.

During the machining operation, surface fibres are damaged and some uneven compression of the wood surface may occur, particularly if surfacing is carried out at too high a moisture content or blunt cutters are used. Surface wetting or cyclic moisture content changes

Timber

resulting from changes in the relative humidity of the surrounding air will cause a recovery of this surface to occur. An early effect of exposure is that the surface becomes somewhat rougher than in the freshly surfaced state.

Many of the extractive chemicals in timber are water-soluble materials which are gradually washed from the surface by rain. Those extractives which are not initially water-soluble are eventually degraded into a soluble form by the action of sunlight. The effect is to remove all the colouring materials from the surface of the timber giving it a bleached appearance. The attractive silver-grey colour of naturally weathered wood is not always achieved in practice, however. On a microscopic scale the wood surface is rough and airborne dirt becomes ingrained to an extent dependent on the degree of exposure and the extent of atmospheric pollution. Also, the protection afforded by eaves retards the loss of colour from these areas and results in a non-uniform appearance.

Because the extractive components of timbers vary, so does the rate at which colour loss occurs. However, these differences are insignificant in comparison with those due to the degree or type of exposure. In the northern hemisphere the south aspect is the most severe as far as colour loss is concerned, followed by the west, east, and north. This reflects the importance of light-induced breakdown of colouring materials on this process.

Weathering

On exposure, unprotected wood gets wet. In response to changes in moisture content resulting from this direct wetting, or changes in its equilibrium moisture content, the wood changes its dimensions. As a result of the organization of the cell walls, the dimensional change along the grain is small enough to be ignored, but across the grain a linear change of up to 5 per cent (depending on species and orientation) may occur in changing from the oven-dry condition to the fibre saturation point. This last level is that at which the interstices of the cell walls are saturated with water but when the cell cavities are still empty. It is about 28–30 per cent moisture content based on the oven-dry weight of the wood. Since wood in service is never dry, the practical limit of linear dimensional movements is about 2–3 per cent.

Short-term moisture content changes affect only the surface layers of a piece of wood since it is not a porous material in the same sense as brick or concrete. Indeed certain timbers are extremely resistant to the penetration of liquids and even with permeable ones, saturation by submersion may take days or even weeks.

The effect of a fluctuating moisture content, and hence changes in dimension against the restraint of the underlying timber, is to repeatedly stress the surface layers. Eventually the fibre-to-fibre bonding fails and splits and checks develop in the surface. At a later stage fibres are lost from the surface, particularly from the softer portions of the growth-rings and the typical fibrous and corrugated appearance of weathered timber develops. Chemical degradation of the wood surface occurs as a result of irradiation with ultra-violet light and this aspect of weathering has been simulated in the laboratory (2–5). A rough fibrous surface is considerably more porous than a freshly prepared one so that the process of weathering accelerates once the surface has been broken down. It is, however, a very slow process. Figures for UK conditions are not available, but an erosion rate of 0·06 mm per year or $\frac{1}{4}$ in per century has been arrived at based on observations on unpainted colonial houses in the Eastern United States (6).

If this were the only aspect of performance to be considered, sacrifice of the surface at this rate would probably be acceptable. The use of correct fastenings, durable timbers, and sufficiently thick sections would enable the use of untreated weatherboarding. This was, in fact, the basis of the weatherboarding on the old colonial houses

Timber

referred to above. Cladding used on modern building is used for its decorative as well as its utilitarian propertes and certain effects of a weathered surface detract from the appearance even if the 'weathered effect' is acceptable. The presence of microchecks, visible checks, and eventually splits in the surface layers of the timber allow water to penetrate into the wood. Gross dimensional changes result, splits enlarge particularly in the vicinity of fastenings and the boards cup and twist with changes in moisture content. Additionally, dirt becomes ingrained in the degraded surface and except under unusual circumstances such as some marine exposures, the silver-grey appearance of weathered timber deteriorates into a dirty grey.

WEATHERING CHARACTERISTICS OF OTHER WOOD-BASED PRODUCTS

Plywood

After timber, plywood is the main form in which wood is used as a cladding material. Although, in general, its weathering behaviour is like that of the timber from which it is made, it has certain characteristics which require special consideration when it is used externally.

Most exterior grade plywoods are made up from peeled veneers. A veneer is a thin sheet (1–5 mm thick except for decorative purposes) produced either by turning a log against a knife which moves inwards at a rate which controls the veneer thickness, or by repeatedly sliding a log against a knife and slicing sheets from it. The peeled veneer or rotary cut veneer necessarily presents a tangential surface whereas sliced veneers may be radial, tangential, or any intermediate angle determined by the decorative effect desired. Whichever process of cutting is used the sheet of wood must be bent over the knife as it is cut. Logs are frequently steamed to soften the lignin and make them more plastic but small splits running along the grain, i.e. parallel with the knife blade, result from this local distortion. With rotary cut veneers, especially those cut from small diameter logs, the tendency to check is increased by the need to produce a flat sheet of veneer by 'un-rolling' a log. The depth and frequency of these lathe checks are influenced by a number of factors not least of which are timber species and veneer thickness.

Veneers are made up into sheets of plywood by spreading them with adhesive and laying them up with the grain direction of alternate veneers at right-angles. The outer or face veneers are spread with adhesive on the inside face only. An odd number are usually used so that the grain of the faces is in the same direction and the board is balanced for strength and dimensional stability. Special plywoods differing from this basic form are manufactured for particular end uses but do not differ significantly from the normal type in their weathering behaviour.

The performance of plywood used for exteriors will depend firstly on the type of adhesive used in its manufacture. For exposure in the unprotected state the bond quality must be of the WBP (weather- and boil-proof) type as specified in BS 1203 (7). Such plywood will remain intact on prolonged exposure but the use of a less-resistant glue-line will result in blistering of the veneers and eventually in complete delamination.

Because the grain direction in plywood is alternated in the veneers and because wood is much more dimensionally stable and strong along the grain, internal restraint in two directions is provided and moisture-induced movements in the plane of the sheet are much reduced compared with solid timber. The combination of the second or crossband veneer under the face veneer and the presence of lathe checks results in the development of fine checks in the face veneer of unprotected plywood exposed to the weather. With a fully exterior plywood this surface checking is of little significance beyond affecting the appearance but it does affect the performance of exterior coatings.

Timber

Blockboards and laminboards (which consist of face and crossband veneers applied to both sides of sheets composed of edge-glued strips of timber or narrow pieces of veneer at right-angles to the plane of the board respectively) are normally produced in interior grades only. Where special-purpose boards are produced for exterior use, the points discussed in relation to plywood generally apply.

Particleboard

Particleboard is another widely used composite of wood and adhesive. There are many combinations of particle size and type but in the main a urea-formaldehyde resin adhesive is used in board manufacture. This type of adhesive is not capable of giving bond strengths which will withstand the stresses of recovery of the particles deformed during pressing and those imposed by dimensional changes on weathering. Development of exterior-grade boards using an adhesive of a type similar to those in WBP plywoods is continuing, but there is as yet little information on their weathering behaviour. Springback of compressed particles does occur resulting in an irregular surface. For some applications this may not be a disadvantage but their long-term performance has yet to be proved.

Fibreboard

Board materials using wood fibres and fibre bundles as their basis are termed fibre-boards and are manufactured in a variety of types, thicknesses, and densities. Only oil-tempered hardboard has been used for exterior work in the untreated state and even here only for low-cost utilitarian buildings or temporary structures.

Under the present UK Building Regulations neither particleboards nor fibreboards are considered to be durable building materials and as such are excluded from exposed use on permanent buildings.

DESIGN CONSIDERATIONS FOR TIMBER CLADDING

Timber cladding is usually an exterior finish applied to a building and only rarely does it perform a significant structural role. The structural material may be timber, masonry, steel, or concrete. Racking resistance to a timber framework may sometimes be afforded by a plywood sheathing which also acts as the external surface but usually the functions of cladding are to exclude the elements and provide a decorative surface. These two considerations are closely linked, particularly so since timber is rarely used without some kind of protective and decorative finish. An exception to this generalization is the field of farm buildings where economy and utility are of greater importance than decorative appeal. To these types of buildings which use unfinished timber the discussion of the weathering of timber which precedes this section applies. In most instances, however, consideration of the performance of timber exteriors involves a combined system of the substrate plus various treatments applied to it for primarily decorative reasons.

It has just been said that cladding must exclude rain, wind, etc., whilst providing an acceptable decorative appearance. It is obvious that it must continue to do so throughout the life of the building and so must be durable. The question of weathering has been considered and concluded to be of significance mainly as far as the decorative aspects are concerned. A further aspect that is of importance from both the protective and decorative viewpoints is that of the resistance of the material to fungal attack. This attack may take the form of superficial mould growth or more deep-seated sap stain, neither of which affect the strength properties of the timber but only detract from its appearance. More seriously it may involve wood-

Timber

rotting fungi which, if unchecked, may cause breakdown of the timber and necessitate replacement.

For attack by any of these agencies it is a pre-requisite that the timber must be above a moisture content of about 20–22 per cent. Such conditions in the timber are very undesirable for other reasons, particularly dimensional stability and durability of applied coatings. Prolonged exposure to these high levels of moisture will give rise to a potential decay hazard. There are two ways of ensuring that decay does not take place and very often they are used in conjunction with each other. The first precaution is to design, construct, finish, and maintain the component so that the moisture content does not reach this critical level. The ways in which this is done will be discussed later and it is good design practice regardless of the decay consideration. On the assumption that design, workmanship, or maintenance may fall below the necessary safety levels, the second combative measure is to use timbers with inherent resistance to fungal decay, or susceptible species that have been given preservative treatment.

Natural durability and preservative treatment

The natural decay resistance of timbers shows a continuous series from highly susceptible or 'perishable. to the 'very durable'. Similarly with preservative treatments it is possible to impart a small amount of resistance to a non-durable timber or to afford it almost indefinite protection, In both instances the choice depends on the economics of the situation.

With building timbers in ground contact, it is easy to define the decay hazard and to specify the level of natural durability or degree of preservative protection required. In this case the probability of the timber exceeding 20 per cent moisture content for long periods is 100 per cent. With other applications of timber, the probability levels are impossible to assess in detail and must of necessity be broad generalizations (8, 9). Under these circumstances, the use of a durable species or specification of a preservative treatment is very much in the nature of an insurance policy taken out against the possible effects of malpractice or neglected maintenance.

Certain levels of protection are required by the Building Regulations but these vary between the Scottish Regulations and those for England and Wales. Basically, hardwoods require no preservative treatment when used as cladding. Softwood timbers require a minimum preservative treatment according to methods prescribed in BS Code of Practice 98 (10). Minimum requirements are also laid down by the National House Builders' Registration Council. In houses built under their scheme, European redwood must receive a minimum dip treatment of one minute in an organic solvent type preservative. Treatment to an equivalent level of protection is required for other non-durable species.

The fact that a relatively low level of preservative treatment is considered, on the basis of present knowledge, to be adequate indicates that the probability of a decay hazard arising in cladding is low. It is nevertheless considered necessary to give a significant treatment to guard against decay should faults allow the moisture content to rise above the danger level.

Timber cladding Timber cladding is available in a very wide variety of profiles but they all have certain features in common. They are of a minimum thickness of about 15 mm and all have some interlocking or overlapping arrangements to prevent water ingress. In addition, many systems make provision for secret fixing of the boards. The economics of conversion of logs into boards and the difficulties of securely fixing wide boards without inducing splitting during seasonal moisture content and dimensional changes, imposes a practical width limit on cladding boards.

Timber

This is usually about 150 mm although an exception is the use of waney-edge boarding in hardwoods such as elm, where the boards may be double this width or more. The width limitation means that in an area of cladding there must be numerous edge joints between the boards. In some respects this is an advantage because the joints are able to accommodate the dimensional movements of the boards. It also enables hidden fixing but does require careful design of the profile of the board to ensure a weathertight cladding.

Board claddings can conveniently be divided into vertical and horizontal types and within each type there is a wide variety of profiles and designs available. More freedom of design is possible with horizontal cladding where overlapping of boards as well as tongue and groove arrangements are possible. With vertical boarding the batten and board or board and batten designs are available as alternatives to the tongue and groove patterns. The design of a profile is dictated primarily by the decorative effect that it is desired to achieve. Vertical pattern boards must not contain butt joints so that more wastage may be involved than with the horizontal types. A selection of the more popular types is illustrated in Fig. 4.3.

Cladding boards are usually given a planed finish. This is always the case when the boards are to be painted but certain of the non-film forming, penetrating finishes perform equally well on sawn surfaces, which may be desired for special effect.

Where horizontal cladding boards are not full length, butt joints are used and located over supports. Both these and the end details where direction is changed or boards abut other materials constitute areas of weakness for rain penetration and must be given special attention. One method is to bed the board ends in mastic and use cover strips to prevent rain ingress, while an alternative solution is to ensure adequate end sealing and detail so that further coats of finish may be applied at times of normal

Horizontal patterns

Rebated Shiplap | Feather edge | Log cabin | Square edge

Vertical patterns

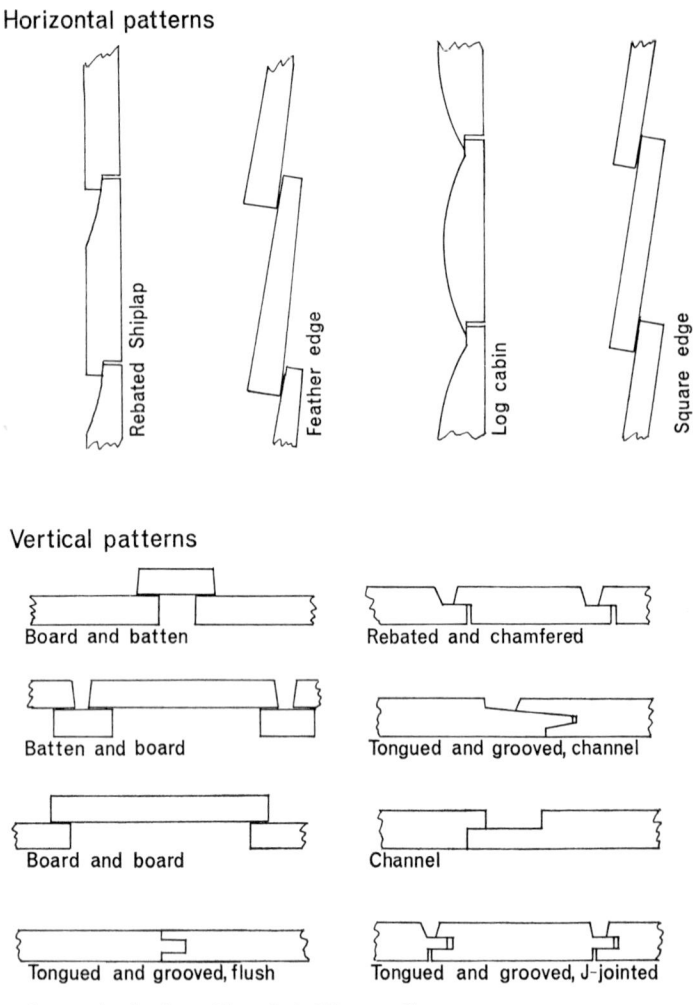

Board and batten | Rebated and chamfered

Batten and board | Tongued and grooved, channel

Board and board | Channel

Tongued and grooved, flush | Tongued and grooved, J-jointed

Fig.4.3. A selection of board cladding profiles.

maintenance. The lower edge of vertical boarding is also vulnerable and must be adequately sealed for good performance. The provision of a drip edge is also desirable to prevent discoloration in this zone.

Timber

Fixings

Fixing of cladding is almost exclusively by nailing. With untreated boards or those treated with a natural finish, wire nails will rust or cause discoloration by reacting with wood extractives. Protected or resistant (e.g. aluminium) nails must be used even when hidden. Boards are fixed over one edge only or at the centre. This is to avoid restraining the movement of the boards which would cause splitting. Pre-drilling of fixing holes with certain timbers, especially denser hardwoods, may be necessary to prevent splitting around the fixings. To provide adequate fixing, the nails used should penetrate the supports to a depth of double the thickness of boards being fixed. Where not hidden, nails should be punched home and stopped to reduce the possibility of their working loose.

The support for timber cladding may be a timber stud framework or sheathing applied to it. Where building materials other than timber form the structural part of the building fixing may be to battens. These battens must be sufficiently heavy to provide adequate fixing and should be spaced at centres suitable for the board thicknesses used.

Timbers

The choice of timber species for cladding is governed largely by cost and secondarily by the technical properties of the timber. The use of natural finishes will further restrict the choice. Assuming an acceptable appearance, or adequate paintability, and cost, a desirable cladding species should possess the characteristics of natural durability or amenability to preservative treatment, ease of fixing, dimensional stability, and freedom from marked tendencies to split and warp. In the UK, western red cedar is the most widely used cladding when a natural timber effect is desired and European redwood when the surface is to be painted. Other timbers which have been recommended as suitable include afzelia, guarea, keruing, African mahogany, sapele, utile, iroko, meranti,

Fig.4.4. Shiplap softwood cladding.

Fig.4.5. Vertical weatherboarding of Western red cedar; no treatment yet applied.

Timber

Douglas fir, and various pines. Many other species are suitable where special decorative effects are required or because local availability makes them competitive in price. The properties of many hardwoods and softwoods and their suitability for various uses are available in *A Handbook of Hardwoods* (11) and *A Handbook of Softwoods* (12).

Plywood cladding

Plywood has the advantage over solid timber of being available in larger sizes, thus reducing the number of joints to be protected from the weather. The most straightforward use as an external building surface is when the functions of sheathing and cladding are combined. Here the plywood is fixed directly to the timber stud framework. The joints between boards are filled with a non-hardening mastic to allow movement and covered with fillets of wood or some alternative material. A plain surface may be used or, alternatively, there are a variety of special effects available. These include simulated random—or standard—width board effects, achieved by grooving the face veneer, and a variety of surface textures. Plywoods for subsequent painting are usually pre-sanded. Those for stain finishes are best left unsanded and slightly rough when surface checking is less noticeable. With plywoods of certain species, where the contrast in hardness between the earlywood and latewood is great, a special 'etched' effect may be achieved by sandblasting the surface or scarifying it with wire brushes to remove some of the softer tissues. This gives the panel a pre-weathered appearance and an interesting texture. Similarly various striated patterns are available which break up the rather powerful grain patterns of certain types of plywood produced from rotary cut veneers.

An alternative way of using plywood to clad buildings is to employ it as infill panels in curtain walling or similar frameworks. For this purpose the plywood is usually glazed in position using mastics to prevent entry of water at the joints. The use of beading to hold the panels in place together with that of non-setting mastics allows a weathertight system of fixing whilst still allowing the small dimensional changes, to which the panels will be subject, to occur. Any of the surface effects described above would be suitable for this usage and in addition it is possible to make use of decorative surface veneers as a special feature as long as their glueing complies with the requirements for exterior durability.

Vapour barriers The importance of excluding rain has been stressed and some of the methods of doing this have been mentioned (flashings, tonguing and grooving, cover mouldings, and mastics). A second potential source of moisture accumulation in the cladding is moisture emanating from inside the building condensing within the timber. The tendency of moisture vapour to migrate from the interior of a building to the outside is not as marked in the UK as in countries with more extreme climates. Nevertheless, there is still the possibility that if precautions were not taken condensation might occur in the cladding in cold weather even here. These precautions are the use of a vapour barrier behind the internal lining to restrict the outward migration of moisture vapour and the positioning of a more permeable breather paper immediately behind the cladding. The function of the breather paper is to prevent the further ingress of wind-driven rain which may penetrate the joints of the cladding whilst still allowing the unrestricted passage of that small amount of water vapour which migrates through the vapour barrier. This combination ensures that moisture vapour does not accumulate in the cavity or structure.

Timber

Fig.4.6. Vertical softwood weatherboarding and plywood infill panels in curtain walling.

Fig.4.7. Vertical pattern softwood cladding on school building.

THE USE OF FINISHES ON EXTERIOR TIMBER

It is usually a requirement of the exterior of a building that it should retain its original appearance throughout the life of the building. Probably this ideal is never achieved even with frequent maintenance but it is a primary reason why timber and other wood cladding products are not frequently used in an untreated state. These materials will not retain anything approaching their original appearance for more than a few weeks when fully exposed to the weather.

Prevention of this deterioration in appearance is brought about by protecting the wood surface from the weathering agencies with a treatment which acts as a barrier to moisture and in many cases to light also (13). These exterior finishes for timber can conveniently be considered as 'natural' finishes on the one hand and heavily pigmented or opaque coatings on the other. There is no clear demarcation between them as some of the natural finishes contain pigments. There are several other ways in which finishes may be classified but none of them are any more precise and since the degree of pigmentation of a finish greatly affects its performance there are practical reasons for considering these two general types separately.

In terms of volume of finish consumed annually for treating timber, far more of the pigmented category, which includes conventional paint, is used. However, since timber is often used for its decorative as well as its functional value, the natural finishes will be considered first. Within the natural finishes, a broad distinction can be made between the film-forming types and those finishes which penetrate the surface leaving little in the way of a surface skin or film.

Natural exterior finishes

The term 'natural' finish is not a particularly appropriate one because any liquid applied to a wood surface will alter its appearance even if the finish is penetrating and colourless. The replacement of air in the cavities of the surface cells has the effect of darkening the surfaces and making it appear redder. The formation of a film on the surface will alter the reflectivity and certain types of finish impart a very high gloss. Such changes from the appearance of freshly planed timber are usually acceptable and indeed are frequently referred to as 'bringing out' the natural appearance of the timber.

The ideal natural finish should:

a. Maintain the original appearance of the timber by protecting the wood surface from moisture and ultra-violet light;

b. Itself be durable and resistant to weathering so allowing long intervals between maintenance periods;

c. Be cheap, easy to apply, and tolerant of malpractice;

d. Be capable of being maintained with a minimum of labour;

e. Be compatible with all wood substrates.

Unfortunately all present-day natural finishes fall short of these requirements in one or more respects. The result is the availability of a large number of different formulations each with their own advantages and disadvantages

Varnishes. These are often produced by reacting a resin with a drying oil and they dry by oxidation, i.e. reacting with oxygen from the air. The drying oils are all of vegetable origin, for example, soya-bean oil, tung oil, linseed oil, and frequently a blend of two or more types are used. Originally the resins were naturally occurring ones such as copal or damar, but current formulations almost always rely on synthetic resins. Alkyd and phenolic resins are those most frequently used.

Fig.4.8. Vertical cladding of Western red cedar, varnish finish.

The relative proportions of drying oil to resin give a varnish its oil length, hence the terms short-, medium-, and long-oil varnishes. The proportions of these two main constituents largely determine the physical properties of the varnish. Resins are hard, brittle materials and short-oil varnishes with high proportions of resin reflect these characteristics. Drying oils, on the other hand, are slow-drying liquids, polymerizing on exposure to the air to give flexible, elastic films. Long-oil varnishes behave similarly, although the presence of the resin gives them tougher films than those of unmodified drying oils. Long-oil varnishes are the most suitable for exterior use.

It has been said that to be effective a finish must protect the wood surface from weathering. Varnishes do this by interposing a more or less transparent film between the weather and the wood surface. This film must remain in position and remain intact if it is to perform its function. It must be well bonded to the wood surface and it must be sufficiently thick to prevent water from penetrating it. In practice no finish is impermeable and both seasonal and shorter-term fluctuations in wood moisture content occur beneath the film as a result both of moisture vapour migrating through the film and from changes via the back, which in practice is frequently unprotected by a finish. On a gross scale these dimensional changes are slight. At a microscopic level they are of greater significance although their actual magnitude is not known. Apart from the fairly straightforward shrinking and swelling of the cell walls there are two other forms of dimensional change which affect the durability of the varnish (or paint) film.

The first of these can be considered to be a defect in the timber and results from incorrect machining of the surface. If machine planing is carried out with the wood at too high a moisture content when the cell walls are more plastic, and particularly if the cutters are blunt, there is a likelihood that the hard latewood zones will resist cutting and will be pushed down into underlying zones of softer earlywood. This fault is especially likely to occur with soft timbers or where there is a great contrast between the physical properties of the early- and latewood zones and where the growth rings meet the surface at an acute angle. On subsequent moisture content changes partial recovery of the compressed earlywood occurs. If a finish film has been applied by the time this occurs considerable stresses are set up in the film at this point and failure by cracking of the film or loss of adhesion may occur. The second type of dimensional change is that which occurs at the junctions between cells of different types whose cell wall constructions and swelling properties differ. The magnitude of these changes is much less than when the machining defect is present and their effect is less well understood. It is probable, however, that when a varnish film does begin to deteriorate the initial failures occur at these discontinuities in the wood structure.

Varnishes are subject to weathering just as is wood (and virtually every other material). They are used to protect timber because the changes which they undergo on weathering are less visible. When first applied they dry quickly from the surface, a reaction aided by metallic soaps termed driers which hasten the oxidation process, but they continue to polymerize and harden for many weeks. They become harder and less flexible as this curing or drying process proceeds. On weathering they are subjected to photochemical breakdown and leaching in much the same way as would be the wood surface which they are protecting. The effect is to reduce still further the film's plasticity (14, 15) so that it is no longer able to accommodate the movement of the underlying wood. The film fails first at those points where the stresses are greatest, for example, the junctions of ray cells and longitudinal cells, or at those points where the film is thinnest and most subject to weathering, for example, the edges of vessels in hardwoods.

This weathering is primarily a surface effect and it follows that the thicker the film, the longer it takes for the embrittlement to reach the wood/varnish interface. Within practical limits, the thicker the varnish film the more durable it will be.

Appearance demands that the thickness of the coating be approximately uniform. To ensure this and to simplify the process of application, the viscosity and thixotropic properties of a varnish are controlled by its formulation. It is necessary to apply several separate coats, allowing each to flow, level, and dry, in order to build up a film of adequate thickness. It is frequently recommended that the first coat should be thinned slightly (10 per cent) to aid its penetration into the wood surface. It is not entirely clear whether this advice is based on experimental facts, practical experience, or intuitive thinking. The use of more solvent will certainly lower the viscosity but since resin and drying oil molecules are too large to penetrate into the cavities of intact cells this may well be irrelevant. A small-scale experiment carried out at the Timber Research and Development Association (TRADA) to assess the influence of thinning the first coat of a three-coat varnish system failed to provide any conclusive evidence as to whether it enhanced durability by promoting better adhesion, or downgraded performance by reducing the contribution to film thickness of the thin first coat.

Although it has been said that the thickness of the film is important in determining the performance and durability of the finish, it is not an easy measurement to make in practice. The conductivity and free film techniques appropriate to non-porous substrates will not work with timber. It is possible to cut transverse sections of films applied to timber (Figs. 4.9–4.12) and to stain them differentially to observe the finish. However, the degree of penetration varies widely and it is usually possible to measure the thickness of only the second and subsequent coats with any degree of accuracy. Even then it is a very time-consuming process. The extent to which the first coat of finish penetrates will depend on the degree of damage to the superficial cells during preparation of the surface, the inclination of the cells to the surface, and the occlusion of the cell cavities by extractive or debris from sanding the surface. In theory, the axis of the cells should run parallel with the surface since boards or veneers are cut longitudinally. In practice, the cells often meet the surface at an angle, which, particularly in the case of timbers having interlocked grain, may be as high as 30°. In such cases, penetration of the finish is great and several coats may be required before the surface is 'sealed' and a surface film is built up. The extreme example of this effect, is, of course, end grain, where the cell cavities meet the surface at right-angles.

Since thin areas in the finish constitute weak areas, a uniform film thickness is desired for maximum durability. A smooth substrate and an absence of sharp edges are necessary. Wet films tend to flow away from edges and a similar effect occurs at irregularities in the surface. The pores of timbers with large vessels should be filled by using a clear filler prior to applying the first coat. Alternatively, pigmented fillers matched to the particular timber can be used. Sanding of the wood surface before finishing helps to

Fig.4.9. Photomicrograph of transverse section of oak treated with two coats of varnish.

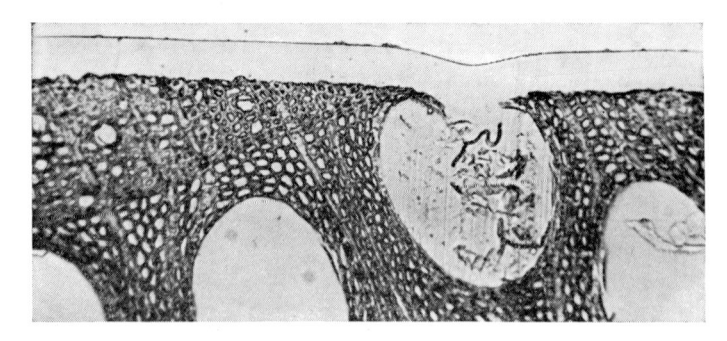

Fig.4.10. Photomicrograph of transverse section of Western red cedar treated with two coats of aluminium wood primer.

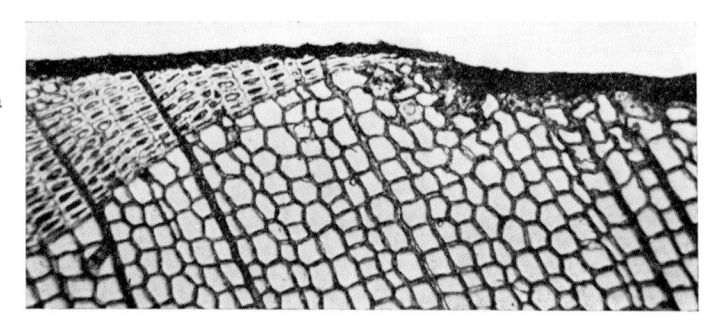

Fig.4.11. Photomicrograph of transverse section of beech treated with two coats of aluminium wood primer.

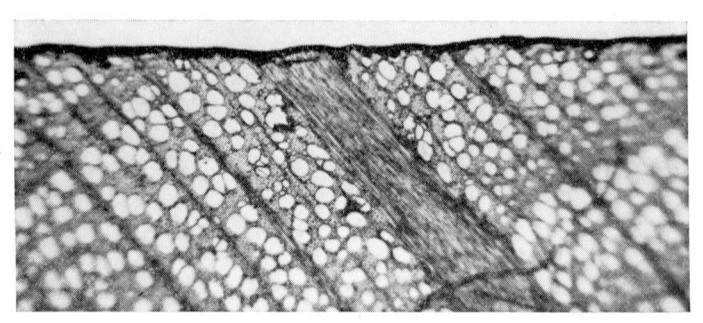

Fig.4.12. Photomicrograph of transverse section of Western red cedar treated with two coats of varnish.

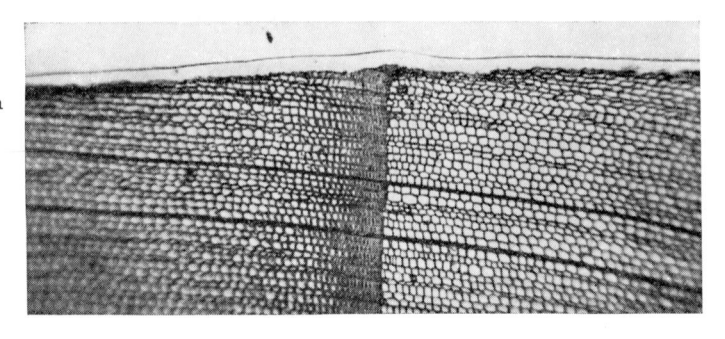

make it more uniform because the fine dust clogs the larger pores. Loose fibres on the surface may project through several coats of finish, causing sites of premature failure. These should be removed by a fine sanding (a process termed denibbing) once the first coat is dry. Besides causing points of low film thickness, these fibres act as wicks and allow water to penetrate the film and so reduce its adhesion to the wood.

Except with very impermeable timbers, the first coat serves only to seal the surface and even out differences in porosity. The build-up of the varnish film really only starts with the second coat and the effect of subsequent coats on durability is additive up to five coats.

During the life of a varnish or other film-forming finish it is subjected to forces which attempt to drag or push it from the surface. The increasing incompatibility as ageing of the film proceeds results in stresses in the adhesion zone. Moisture vapour pressures set up behind the film tend to induce loss of adhesion and blistering. With modern finish formulations, blistering usually indicates that there is an excessive build-up of moisture behind the film due either to the wetting of the back of the boarding or to migration of moisture vapour from the inside of the building. Prevention of leakage will cure the first cause and the use of a vapour barrier or more impermeable coating on the inside, the second.

Despite the number of different formulations possible within the alkyd and phenolic varnish types, it is possible to generalize on their characteristics. An alkyd varnish tends to be clear or only slightly yellow in contrast to the phenolics, whose darker resins impart to them a distinct yellow tint. On the other hand the phenolic varnishes are more water-resistant than the alkyd types and therefore possibly somewhat more durable. The durability of a particular varnish formulation is not yet predictable since all the factors affecting this property are not fully appreciated or understood.

Synthetic resinous clear finishes. Other film-forming or resinous clear finishes are similar in behaviour to the traditional varnishes but are of different chemical types. They are synthetic polymers which harden by the chemical reaction of their components. The drawbacks of traditional varnishes and the rising cost and increasing scarcity of the natural raw materials from which most durable types were made, stimulated the search for alternative and better products. Products of several chemical types have in turn been reputed to be the long-sought-after breakthrough in clear finishes for timber. None has so far achieved this and, by and large, finishes based on epoxy, urea formaldehyde, and isocyanate resins have not proved superior to the true varnishes. With the exception of formulations of the latter type, there are relatively few commercial finishes of these types on the market today inthe UK.

Finishes based on isocyanates, the polyurethane clear finishes, were introduced as two-pack materials, i.e. the reactants were mixed together immediately before use and began to set immediately. Thus their pot life was limited and excess finish could not be used. The inconvenience of the two-pack system was not shared by a later development, the moisture-curing polyurethane. Although chemically similar to the previous type, the moisture-curing type is a single-pack material which is stable in the absence of moisture. Moisture vapour from the atmosphere or from the substrate initiates the curing or setting reaction which is rapid in comparison with the oxidation drying of varnishes. Both types of polyurethane varnish cure to give strong, tough films with a high degree of chemical resistance. This resistance is advantageous in certain situations but may cause difficulties if the interval between the application of successive coats is more than 24 hours. Excessive cross-linking of the molecules in the film leaves it unaffected by the solvents in the wet film and the greater the delay, the greater the possibility of low intercoat adhesion and eventual failure by delamination of

Fig.4.13. Exposure panel of Western red cedar illustrating edge failure of two pack polyurethane finish after natural weathering.

the various coats. This characteristic poses a problem when a weathered film is to be maintained. A careful cleaning and abrading of the weathered surface is essential to improve the adhesion. A further disadvantage of this type of finish for exterior use on large areas is also related to the strength of the film. As this weathers it contracts. The high strength of the film coupled with the low edge-build characteristics of this type of finish result in a splitting of the film at sharp edges (Fig. 4.13). It will be seen from this illustration that this occurs almost irrespective of the number of coats. This type of failure is a contradiction to the generalization that a greater number of coats will give greater durability.

It also is very difficult to rectify this edge failure on maintenance. The films are resistant to chemical paint strippers so that complete removal is tedious, while treatment by feathering the damaged edge of the film followed by a bringing forward of the peeled area is unlikely to be much more than a temporary repair. The old film continues to contract and pull away from the edge, splitting the new coats in the process.

A third type of polyurethane varnish, the oil-modified polyurethane, has recently been introduced. In formulation it has much in common with traditional varnishes but is harder and more resistant. It is rather early yet to say whether these formulations suffer the disadvantages of the previous types. It seems likely from those exposure trials which have been carried out that they behave more like the alkyd and phenolic varnishes than like the two-pack and moisture-curing polyurethanes.

All types of varnishes and other clear finishes are transparent to ultra-violet light to a greater or lesser extent. The synthetic finishes are generally more transparent unless special ultra-violet-absorbing additives are included. It is this transparency which distinguishes the mode of deterioration and failure of a clear finish from that of a pigmented one. When much of the ultra-violet energy is dissipated by the film, the photochemical degradation of the wood surface is reduced. With the more transparent finishes, the film may undergo relatively little weathering but chemical and physical changes, leading to a disruption of the wood structure and loss of adhesion, will occur. In an extreme case much of the failure may occur in the surface layers of the wood rather than in the finish.

With a clear film coating, failure of the integrity of the film will eventually occur unless preventive maintenance is practised. Initially, these breaks in the film will be small or they may involve extensive edge breakdown. Wherever they occur water is then able to enter and more rapid deterioration sets in. The water itself causes discoloration and both chemical and biological changes continue this effect. The biological agencies involved have not been rigorously studied but probably include soft rot fungi and bacteria as well as mould fungi (16). Areas of discoloration spread, film peels back from the splits, and appearance deteriorates rapidly.

Unlike a paint, maintenance coats of varnish will not hide this discoloration. In the same way that initial preparation of a wood surface is more critical for a varnish than a paint, maintenance must be either anticipatory or promptly executed on the onset of film breakdown if difficult and costly rectification is to be avoided.

Timber

The actual durability of a resinous clear finish will depend not only on the formulation of the product but also on factors such as the substrate, exposure, number of coats, time of exposure, care of surface preparation and conditions at the time of application. Under normal practical conditions, a life of a three-coat system of three years is probably the maximum that can be expected even with the best formulations and in many cases it may be considerably less. This level of durability, coupled with the cost of initial application and prompt regular maintenance, leaves film-forming clear finishes far from the ideal natural finish for timber.

Penetrating natural finishes. Drying oils were the forerunners of current penetrating finishes. The use of linseed oil on external timber is still practised today, usually with unfortunate results. Drying oils, even when modified by heat treatment and the incorporation of driers, are slow-drying finishes by nature. Linseed oil (even the boiled variety) may take months to dry on a north aspect. This characteristic, coupled with its low water-repellency, leads to dirt accumulation which is impossible to remove satisfactorily without a complete resurfacing of the wood. The oiled surface may also be prone to mould growth under damp conditions and the situation is aggravated.

Teak oils (drying oils modified with resin) whilst not suffering from the disadvantages of linseed oil have a very limited exterior durability and are only suitable where frequent reapplications can be ensured.

Attempts to overcome the limitations of straight linseed oil led to the development of a 'natural finish for wood' (17) at the Forest Products Laboratory in Madison, Wisconsin. This formulation has become known as the 'Madison' formula and, either in its original form or with modifications, forms the basis of many current commercial products. It consists of boiled linseed oil, turpentine, or paint thinner, 2·6 per cent by weight of paraffin wax,

5 per cent by weight of pentachlorophenol (a preservative), an anti-bloom agent, and iron oxide pigments to add a limited amount of colour. The viscosity of the material is low and when applied to most timbers it penetrates, leaving little residue on the surface.

The addition of paraffin wax to the formulation confers water-repellency on the finished surface since the exposed surfaces of the cells are coated with a thin film of this highly hydrophobic substance. Whereas a film-forming finish prevents weathering by excluding water with an impermeable barrier, the so-called water-repellent finishes do so by making the surface of the timber shed liquid water (Fig. 4.14). They are not resistant to the penetration of water vapour, so that water lying on a horizontal surface will gradually be absorbed as vapour. This ability to transmit water vapour, or, more popularly, to breathe, is an advantage in that it prevents blistering of the finish.

As a result of modifications to the Madison formula and the development of entirely new products based on the same principle, there exists today a variety of products which, for convenience, are grouped under the term

Fig.4.14. Exposure panels of Western red cedar treated with water repellant preservation stains.

'water-repellent preservative stains'. They vary from products which are basically organic solvent type preservatives to which water-repellents and stains have been added, to highly developed finishes based on synthetic resins. Some are capable of being applied by dipping; most are suitable for spray or brush application and others are of a mayonnaise consistency. As might be expected, they also vary in price and performance.

The optimum level of water-repellency to be aimed at is still in doubt. Too low a level will result in low weathering resistance and too high a level in dirt accumulation as a result of interference with the cleaning action of rain. Besides paraffin wax, water-repellency is achieved by waxes other than paraffin wax, silicones, and synthetic resins in the various formulations. In a recent comparative trial (TRADA, unpublished results) all the most durable materials had high levels of water-repellency as measured by the TRADA method (18).

Most of the modifications to the original Madison formulation have been directed towards reducing the tendency for dirt to accumulate on the finish and cause darkening. The incorporation of a small proportion of synthetic resin to speed drying and seal the surface has been found desirable in many cases. Some formulations even produce a somewhat glossy effect but the film thickness is very small compared with that of a varnish.

A number of products contain little or no pigmentation in spite of the term 'stain' in the generic title. These products are designed for the intial treatment of cladding and their durability is generally inferior to that of their pigmented counterparts. They do, however, come closest to maintaining the natural appearance of the timber, at least for short periods. Most natural versions of water-repellent preservative stain finishes are pigmented to a varying degree with brown/yellow/red pigments to simulate the appearance of timber. Other formulations may use pigments of other colours for special effects. Some of

Fig.4.15. Vertical softwood cladding treated with a dark, water-resistant preservative stain.

Fig. 4.16: Softwood boarding, treated with black water repellant preservative stain, used as cladding and open screening to access stairs of four-storey maisonettes. Plywood infill panels are also illustrated.

Timber

these approach paints in appearance as the level of pigmentation is so high that grain pattern of the timber is hidden.

Water-repellent preservative stains do not fail by cracking or peeling since there is little or no surface film. On exposure to the weather, they gradually erode from the surface. The level of water-repellency drops during exposure and both pigment and binder are lost. This process usually occurs uniformly over the surface except where this is sheltered from rain and sunlight. The result is a gradual change in appearance rather than an obvious deterioration and the decision as to when retreatment is necessary is a subjective one. Since retreatment adds further colour to the surface, it is an advantage if the finish bleaches on weathering rather than darkens. With finishes which darken, repeated applications will result in a steadily darker appearance. Cleaning with a detergent solution to remove superficial dirt before retreatment is to be recommended. All traces of detergent must be removed and the timber allowed to dry before application of the maintenance coats. Even with those formulations which do not darken, periodic cleaning before retreatment may be necessary particularly if rain splash results in dirt accumulation on the footings of the cladding.

This ease of retreatment is one of the main factors in favour of this type of treatment for natural cladding and it contrasts sharply with the procedures necessary to maintain a resinous clear finish in good condition. Initial treatment is similarly less demanding. The surface should be clean and dry but there are no other requirements. All timbers are suitable and surfaces may be planed, sawn, or textured in the case of plywoods, for which these finishes are also appropriate. With light-coloured, non-durable timbers, a prior treatment with preservative to prevent staining is desirable since the preservative in many of the formulations is to prevent mould growth on the finish rather than to confer durability on the timber.

It is normal to apply two coats both initially and on retreatment. Some products suggest one coat, others three, and manufacturers recommendations should be followed. An average 'life' of this type of finish is even more difficult to predict than for a varnish, but two to three years appears reasonable for good formulations.

Except for the clear materials with their lower durability, the water-repellent preservative stains modify the initial appearance of the timber to a greater extent than the varnishes and in a way which may be less acceptable. To offset this, however, are the advantages of ease of application and maintenance, the choice of numerous special effects, and the less-critical nature of the time of retreatment.

Opaque coatings

Where it is not desired to retain a natural appearance, it is possible to use coatings which are far more durable than any of the natural finishes. These paints are essentially varnishes (the vehicle or binder) into which are incorporated pigment particles which give them their opacity and durability. Many factors affecting their performance are those which influence the clear finishes and have already been discussed. Ways in which the two types differ are frequently a result of the pigmentation of a paint and the greater freedom of pretreatment and formulation which this allows.

The multiplicity of chemical types found with the clear coatings does not exist with decorative exterior paints. Almost all current commercial products are based on an alkyd resin. The exception to this is found in the case of primers, the first coat of a paint system which has no real counterpart in a clear finish system.

In the UK an exterior paint system on wood is traditionally a three-coat system consisting of primer, undercoat, and finishing or top coat. For enhanced durability either the undercoat or the finish coat may be doubled

Timber 173

Fig.4.17. Variety of timber usage and finishes on exterior of school building.

Fig.4.18. Combined use of horizontal and vertical weatherboarding on rural school.

up (12). The paint for each of these three coats is formulated with specific properties in view.

Primers. A primer is the first material to be applied. The primer coat forms the link between the timber and subsequent coats and must therefore have good adhesion to the substrate and afford a good surface for application of the undercoat. It serves to seal the wood surface to prevent excessive migration of the liquid part of the undercoat. It is usually pigmented to modify the appearance of the timber and contribute to the opacity of the finish and must possess a reasonable degree of water-resistance and durability. It is perhaps less prevalent with cladding than with other items of exterior joinery but preprimed items are frequently stored on a building site under poor conditions where weathering and wetting occur. In traditional wet construction, these items are then built into wet masonry with only the primer protecting the timber. Further weathering may occur as it is frequently much later that the undercoat and top coat are applied.

Lead-based or pink primer to BS 2521 is the material by which other primers are judged but one which has been almost completely superseded on grounds of cost and toxicity. It is a product with high water- and weathering-resistance, tolerant of substrate, with good filling properties, and amenable to coating with different undercoat and top-coat formulations. Its widespread use was dropped in favour of lower-priced leadless oil primers which unfortunately range from products with almost the performance of the lead type to inferior products yielding weak coats with low weather and water-resistance. Because there is this range of performance and in the absence of a British Standard dealing with this category of primers the false economy of using the cheaper products of this type should be avoided.

The more recently developed emulsion primers have several advantages, particularly convenience in use.

They are water thinnable and equipment can be cleaned with water. In addition they apply easily and dry rapidly allowing undercoats to be applied within a few hours. The pigments used are non-toxic and the solvents non-flammable, leading to greater safety in use. Their weathering-resistance (particularly the acrylic types) is equivalent to that of the lead-based primers but unlike the latter type, they result in a permeable film with little water-resistance. Although they will not satisfactorily exclude water when used alone, they are claimed to be largely self-knotting as they prevent resin exudation more than most primers.

The final category of primer, aluminium primer, can be even more effective at sealing back resin and other possible exudations. It is also highly resistant to water penetration. It is frequently used over oily or resinous timbers and the correct type can be used on timber which has previously been treated with a tar oil preservative such as creosote. It is recommended as the primer for Douglas fir plywood and for end-sealing boards.

Undercoats and finish coats. The past decade has seen a radical change in the types of paint used for the protection and decoration of wood. There has been an upsurge in the use of the synthetic resin alkyd paints at the expense of the older oil types which are now rarely found. The newer paints give tougher films than the softer, slow-drying oil types and the mode of failure is less by erosion of the surface and more by breakdown of the film followed by checking and cracking at points of stress. The new formulations are inherently very durable and a well-applied four-coat system is capable of giving ten years service before maintenance becomes necessary. Unfortunately this life is rarely achieved in practice as a result of failure to take adequate care with surface preparation and application.

A newer trend is the use of water based undercoat and

finish paints using acrylic resin or acrylic and polyvinyl chloride copolymers. Technical difficulties of producing an exterior gloss emulsion paint delayed the introduction of this type of finish. Semi-gloss and flat paints used externally in our climate tend to soil rather quickly and their durability is usually lower than the equivalent gloss finish. Gloss finishes are also more easily cleaned should they become dirty.

Weathering of paints. The same processes of degradation affect paints as wood or natural finishes. A fundamental difference with an opaque coating is that an insignificant amount of light energy reaches the wood surface so that this remains virtually unaltered as long as the integrity of the film is maintained. A loss of gloss is the first visible result of weathering and it is caused by the photochemical degradation of the paint binder and the leaching of components from the surface of the paint film. Continued attack by light and water result in a gradual embrittlement of the paint film and a loss in strength. The underlying wood continues to respond to changes in moisture content. A fresh paint film similarly undergoes dimensional changes with changes in moisture content and indeed these may be more pronounced than those of the substrate. As the paint film embrittles, it becomes increasingly incompatible with the wood until it is eventually stressed to failure. Initially these fissures are on a microscopic level, but with continued weathering, the microchecks in the film enlarge to become visible cracks. The loss in protective power of the film leads to rapid water penetration and a loss of adhesion of the film around the failures. These areas of adhesive failure eventually join up and the paint film flakes off.

For maximum durability the paint film must behave integrally, not as a series of distinct layers. This means that compatibility and adhesion between the coats must be good. The former aspect is assured by following manufacturers' recommendations, the latter by ensuring that the previous film is not contaminated or excessively weathered. The success of a system also depends on the adhesion between the wood and the paint film and this is why the primer is so important since it links two dissimilar materials. Adhesion of the primer is partly chemical (particularly in the early stages of paint life) and partly mechanical resulting from the penetration of the finish into the wood elements. A wood surface contaminated by dirt, dust, oil, grease, natural exudations, chemicals, or water will not provide a satisfactory adhesive bond with a primer.

Of these contaminants, excessive water is probably responsible for the majority of failures to achieve anything like the maximum potential durability from a paint system. The moisture content of the wood should optimally be that at which it will settle down in service so that there is no prestressing of the applied film. In any case it should not exceed about 18 per cent. Various additives have been produced which reputedly allow one to apply paint to wet timber but it is significant that these are not usually accepted by the major paint manufacturers. The presence of moisture beneath a paint or varnish film greatly impairs its adhesion and durability. End grain is particularly susceptible to water penetration not only because it is more permeable than side grain but also because it is difficult to seal with conventional finishes. Particular attention to its sealing must be paid at joints.

The provision of a uniform film of adequate thickness is also important for durability. A wider variety of grain fillers is available when the appearance of the wood surface is of no importance and these should be used on coarse-grained timbers to provide a smooth surface. A smooth sanded surface and freedom from sharp edges which can initiate failure are also advantageous in obtaining the maximum life from a paint coating. A further pretreatment which must be carried out on resinous timbers with knots is the application of a seal or knotting compound. This may be shellac or aluminium sealer and is applied locally over the knot or area of exudation. It is unrealistic to try to seal large resinous knots or pitch streaks and if timber containing these must be painted, it is advisable to cut out the defects and plug the holes with sound timber. Timbers of a naturally resinous nature, for example, keruing, gurgun, agba, are best primed with an aluminium wood primer to prevent or reduce the amount of exudation.

Various properties of a timber affect the durability of paint or varnish applied to it (20–2). These include texture, earlywood/latewood contrast, dimensional stability, and extractive content. A detailed discussion of these can be found in the references cited.

As discussed earlier, preservation treatment of exterior timber is now more worthwhile as labour and material costs for maintenance rise. Although there is little likelihood of a chemical incompatibility between the preservative agents and the finish system (copper napthenate is a possible exception) certain elementary precautions must be taken. The most likely cause of trouble and premature failure is due to the trapping of solvents, used to carry the preservative chemicals, behind the film. These then may blister the film, cause retarded drying or other drying defects, and cause discoloration of the film as they migrate through it carrying colouring materials, such as resins, with them. The precaution necessary to avoid this type of trouble is to allow sufficient interval between the application of the preservative and the paint to allow these excess solvents to evaporate. Since the required period depends on the type of preservative and its formulation, the advice of manufacturers should be sought in particular cases. The same is true for the choice of compatible preservative and primer systems since with the present state of knowledge it is not possible to generalize.

Some of the organic solvent preservative formulation

currently in use contain water-repellent additives of the wax or resin type. They are usually called water-repellent pretreatments. Their object is to supplement the water-resisting properties of the primer and thus give the benefits of a more dimensionally stable wood surface and a second line of defence against water penetration should the paint system fail. Since this type of product is still being evolved and formulations are changing, it is desirable to check that the paint system is compatible with a particular pretreatment. Certain of these materials may contain small amounts of stain or pigment and are sometimes described as water-repellent primers. Until further evidence is available it is probably wise not to regard them as equivalent to conventional primers since their film-forming ability is low. An alternative way of combining the dual functions of primer and wood preservative is to include preservative chemicals in a primer formulation. Again, the performance of these products is in doubt, in this case because the penetration of the preservative into the wood is considered to be inadequate.

Methods of test for finishes have commanded a great deal of research attention over the years, but there are still many difficulties associated with getting quick performance indices on products with an expected life of many years. Controversies still rage over whether accelerated methods of test (weatherometer trials) are appropriate to timber and wood-based substrates and if they are, which particular system should be used. While it is reasonably well established that certain aspects such as chalking of a pigmented film can be simulated (23), the process of natural weathering is so complex, the variables so many, and the interactions so involved that it is likely that for over-all assessment of wood finishes, natural exposure will be relied upon for many years yet. This is not to say that natural exposure is a foolproof method. Weather varies geographically and from year to year and the conditions prevailing when test panels are first exposed has a significant effect on their subsequent behaviour. It is also a slow process in spite of accelerations achieved by special design of test panels, exposure at 45° facing the sun or the use of sun-following exposure racks.

Whatever the method of test, it is rarely possible to simulate all the variables introduced by on-site application of wood finishes (24–5). Temperatures, environmental conditions during application, and workmanship are only three of the factors likely to be more favourable in product evaluation testing than on site.

Factory finishing. Since the way in which a finish is applied has a considerable effect on its performance in service, application under conditions which can be controlled is likely to give films of greater durability. Herein lie the advantages of factory finishing. Although complete prefinishing has only been tried on an experimental scale in this country, the use of this method of applying primers and undercoats is increasing. Specially formulated paints for spray, dip, or mechanical coating are used giving controlled application and rapid drying for compatibility with production-line processing. The next few years will probably see considerable development in this field, and already techniques such as electrostatic spraying and irradiation curing of coatings are being investigated.

Site work is so often plagued by inclement weather that recommended precautions for the application of finishes are largely ignored. Other building practices tend to encourage poor workmanship, the skimping of surface filling, knotting, and stopping operations so that often only a small proportion of the potential durability is achieved. Site application is also almost entirely by brush and, although when done thoroughly this is a good method of application, it is one open to abuse.

Integral finishes. Conventional wood finishes are applied wet and bond themselves to the wood surface as they dry.

As can be seen from the mode of failure of modern finishes, this adhesion is critical as it resists the stresses imposed by the physical incompatibility of coating material and wood which increase as the former component ages. This adhesion is susceptible to the presence of moisture and is stressed by moisture-induced changes in the dimensions of the wood structure. Surface stabilization by resin impregnation or by replacing the natural wood surface with a resin-impregnated overlay has the effect of greatly prolonging the useful life of a finish coating (15). This is particularly true in the case of plywoods where surface checking makes finishing with conventional paints and varnishes only a short-term solution. In these cases, resin overlays will improve the characteristics of the substrate to such an extent that paint performance is equivalent to that on the best timbers. Other forms of surface stabilization (acetylation) are also effective in improving paint durability (15).

An alternative method of prefinishing is to use film overlays which are themselves decorative. Perhaps the most promising of these at the present time is a system based on pigmented polyvinyl fluoride film bonded to a timber or plywood substrate. The durability of this system is claimed to be equivalent to the life of the building. Further developments in this field can be anticipated with the development and evaluation of improved plastics.

CONCLUSIONS

From the previous discussion it will have been seen that a discussion of the performance of timber and wood-based materials as claddings for buildings involved consideration of the performance of a whole range of surface and other treatments applied to timber. These treatments protect and decorate the wood surface and enhance the versatility of this basic material. It has not been possible to go into detail on all forms of wood products used as exterior surfacings; shingles and shakes have not been mentioned for instance. Nor has it been possible to deal with the numerous different types of speciality finishing materials which are available or some of the other considerations such as fire performance which influence the use of wood externally. Most of the remarks that have been made about cladding and infill panels apply to other items of exterior joinery such as eaves and soffit boards. Timber windows can also be included in this list although there are special problems of design and construction here associated with the need to prevent water entry through vulnerable joints and glazing seals.

Development of new finishing materials is taking place continuously and is a process that has been continuing for many years. If past experience is reliable, it seems unlikely that any major breakthrough in the use of finishes to improve the performance of timber and wood-based cladding materials will be made. Rather the situation is likely to be one of modest progress through the development of new materials and improving the techniques and control of applying these and existing materials. In practice there is much left to be desired in the way finishes are applied. Industrialized prefinishing is one way in which the situation can be improved and increasing use of this system is likely to be made as experience with handling prefinished units and components is gained.

Research into the technology of timber exposed to the weather and of treatments applied to it to prolong its service life and enhance its appearance is carried out in the UK by four research organizations which maintain close formal and personal contact to ensure co-ordination of effort. These organizations are the Building Research Station, the Forest Products Research Laboratory, the Paint Research Association, and the Timber Research and Development Association. From them can be obtained advice on specific problems and publications dealing with research and practical developments in this field.

REFERENCES

1. WISE, L. E., and JAHN, E. C. (1952). *Wood chemistry*, vols I and II. p. 1343 (New York: Reinhold).
2. MINIUTTI, V. P. (1965). 'Microscale changes in cell structure of softwood surfaces during weathering', *Official Digest*, **37**, 692–7.
3. —— (1964). 'Microscale changes in cell structure at softwood surfaces during weathering', *Forest Products J.*, **14**, 571–6.
4. —— (1967) 'Microscopic observations of ultraviolet irradiated and weathered softwood surfaces and clear coatings', *US Forest Serv. Research Paper*, FPL 74, p. 32.
5. WEBB, D. A., and SULLIVAN, J. D. (1964). 'Surface effect of light and water on wood', *Forest Products J.*, **14**, 531–4.
6. US DEPARTMENT OF AGRICULTURE (1955). 'Wood handbook', *US Dept. Agriculture Handbook*, no. 72, p. 528.
7. BRITISH STANDARDS INSTITUTION (1963). *Specification for synthetic resin adhesives for plywood*, BS 1203, p. 15 (London: BSI).
8. BRITISH WOOD PRESERVING ASSOCIATION and TIMBER RESEARCH AND DEVELOPMENT ASSOCIATION (1957). *Timber preservation*, p. 89.
9. TIMBER RESEARCH AND DEVELOPMENT ASSOCIATION and BRITISH WOOD PRESERVING ASSOCIATION (1967). *Safeguarding timber in industrialised buildings*, p. 4.
10. BRITISH STANDARDS INSTITUTION (1964). *Preservation treatments for constructional timber*, CP 98 p. 20 (London: BSI).
11. *A handbook of hardwoods* (1956). p. 269 (London: HMSO).
12. *A handbook of softwoods* (1957), p. 73 (London: HMSO).
13. TIMBER RESEARCH AND DEVELOPMENT ASSOCIATION (1966). *Maintaining timber exposed to the weather*, p. 4 (High Wycombe: TRADA).
14. VAN LOON, J. (1966), 'The interaction between paint and substrate', *J. of Oil and Colour Chemists' Assoc.*, **49**, 844–67.
15. TANHOW, H., SOUTHERLAND, C. F., SEEBORG, R. M., and KALNINS, M. A. (1966). 'Surface characteristics of wood as they affect durability of finishes', *US Forest Serv. Research Paper*, FPL 57, p. 60.
16. DUNCAN, C. G. (1963). 'Role of micro-organisms in weathering of wood and degradation of exterior finishes', *Official Digest*, 35, 1003–12.
17. 'Forest products laboratory natural finish' (1964). *US Forest Serv. Research Note*, FPL 046, p. 5.
18. HILL, R. R., and SHARPHOUSE, R. P. (1969). 'TRADA Tentative Standard Test for water repellency of finishes', *Research Paper*, WT/RR/2 (High Wycombe: TRADA).
19. BUILDING RESEARCH STATION (1969). 'Painting woodwork', *BRS Digest*, no. 106, p. 8 (London: HMSO).
20. BROWNE, F. L. (1951). 'Wood properties that affect paint performance', *US Dept of Agriculture, Forest Products Lab. Report*, R1053, p. 23.
21. JONES, K. L. (1966). 'The varnish holding properties of timber', *J. of Oil and Colour Chemists' Assoc.*, **49**, 314–39.

22. Gray, V. R. (1961). 'The wetting, adhesion and penetration of surface coatings on wood', ibid., **44,** 756–86.
23. Bullett, T. R. (1968). 'Durability of paint', ibid., **51,** 894–902.
24. French, E. L. (1968). 'The problem of the factory and site application of decorative finishes for industrialised building', ibid., **51,** 723–9.
25. Seavell, A. J. (1963). 'The durability of paint systems; standard laboratory testing compared with painting time schedules on the building site', ibid., **46,** 791.

Chapter 5 METALS

W. D. HOFF

BSc, PhD, AInstP
*Lecturer, Department of Building,
The University of Manchester Institute of
Science and Technology*

INTRODUCTION

Several metals have a history of architectural application extending over thousands of years, and the large-scale use of metals on building surfaces similarly has a long tradition. Today the demands of technology have resulted in the development and production of many metals and alloys covering an extensive range of well-defined properties. Many ferrous and non-ferrous alloys have found architectural application and the present-day use of metals on buildings extends far beyond the traditional uses of sheet-metal roofing and cladding.

Metals are present on the external surfaces of buildings in a wide range of forms from small fastenings to large structural members, from detailed coverings and flashings to large areas of cladding and roofing. Such a range of applications involves widely differing requirements and the problem in building design is one of selection of suitable alloys which show a satisfactory performance in the chosen application. There are two basic properties required of any external building component. Firstly, the component must maintain its structural integrity over a long period of time, and secondly the component must be aesthetically satisfactory. In all proposed applications of metals on the external surfaces of buildings the general climatic and atmospheric environment of the building should be considered in order to decide on the probable performance of selected metals in the general environment of the building, and secondly, the details of the design must be considered to determine if any problems of interaction could occur between the metal and the surrounding materials and components.

Fundamental to the deterioration and weathering of metals is the phenomenon of corrosion. All metals tend to revert to their natural ores on prolonged exposure to the atmosphere, but in many cases this process is self-stifling, the products of corrosion forming a protective film over the metal surface. In other cases the corrosion takes place at a uniform rate and the metal is gradually destroyed. This latter form of corrosion can be tolerated if a satisfactory thickness of metal has been allowed at the design stage. Otherwise protective measures need to be adopted. Despite these effects the metals used on buildings are inherently durable, and provided that major hazards are avoided a long service life can be expected.

The requirements for satisfactory performance of metals on buildings are the selection of suitable metals for given conditions, the correct detailing of design to avoid corrosion hazards, and the carrying out of suitable maintenance if necessary.

FUNDAMENTAL CONSIDERATIONS

In practice the processes of corrosion can be complex, principally because the corrosion of a metal is affected by minor changes in environment, by variations in com-

Metals

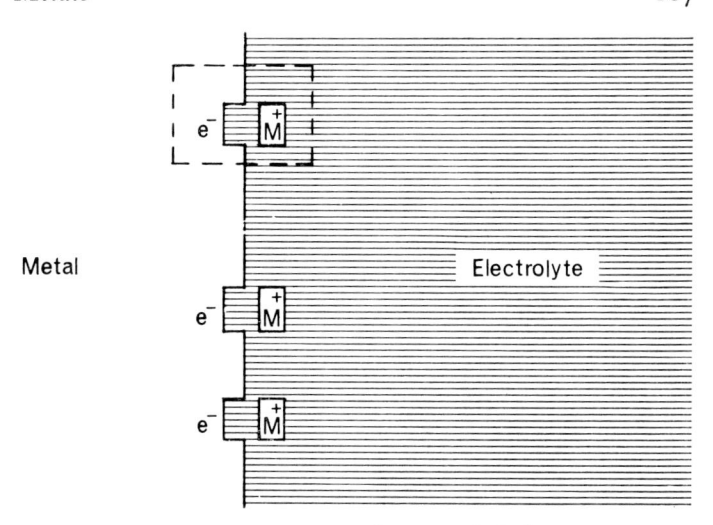

Fig.5.1. Diagrammatic representation of process of solution, e.g.
$M \rightleftharpoons M^+ + e^-$. The negative charge tends to prevent ions from migrating away from the metal surface and impedes progress of the reaction.

position, and by contact with other materials. The basic principles of corrosion are well defined and serve as an accurate guide to those precautions that need to be taken in a practical situation.

The corrosion of metals is basically electrochemical in nature and takes place in the presence of an electrolyte. In this context an electrolyte is a solution containing ions; water, for example, contains hydrogen H^+ and hydroxyl OH^- ions. Pure water is not a good electrolyte since it contains relatively few ions, but the presence of small amounts of salts or of acids or of alkalis increases the number of ions considerably.

When a metal is placed in an electrolyte it tends to dissolve according to a reaction of the form:

$$M \rightleftharpoons M^{n+} + ne^-. \qquad [1]$$

Thus the metal atoms go into solution as positively charged metallic ions leaving an excess of negatively charged electrons in the metal (Fig. 5.1). The build-up of negative

Table 5.1   *Electrochemical series of some of the pure metals.*

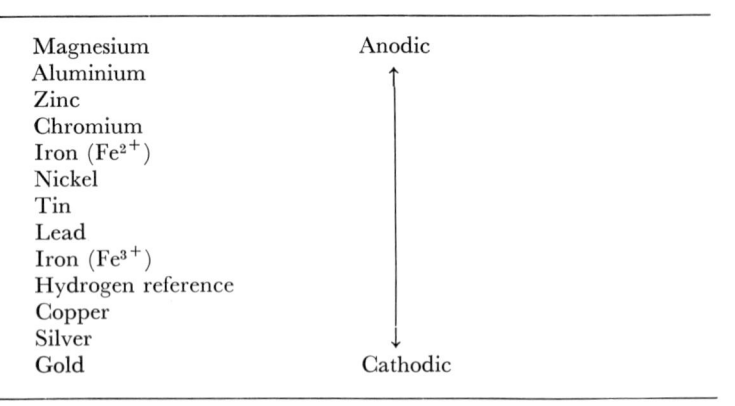

Magnesium	Anodic
Aluminium	↑
Zinc	
Chromium	
Iron (Fe^{2+})	
Nickel	
Tin	
Lead	
Iron (Fe^{3+})	
Hydrogen reference	
Copper	
Silver	↓
Gold	Cathodic

(This series refers to the pure metals uncontaminated by any surface films; a galvanic series (q.v.) should be consulted when a practical situation is being considered.)

charge on the surface of the metal (resulting from the presence of electrons) serves to attract the metal ions back to the surface of the metal and tends, therefore, to inhibit the progress of the reaction. The progress of the reaction depends upon the nature of the metal and the nature of the electrolyte. The tendency of a metal to dissolve in a particular electrolyte is known as the 'solution pressure'.

The above example indicates the way in which the rate of corrosion of a metal may be increased or decreased. If the reaction expressed in equation [1] above is driven to the right, then more ions will go into solution and corrosion will increase. If the reaction is driven to the left the corrosion will be retarded. Therefore, if the metal under consideration is in some way caused to loose the electrons produced as a result of the ionization reaction, then the corrosion of that metal will continue. If it is connected to a source of electrons, the corrosion process will be retarded because the reaction [1] will be forced to the left.

The tendency of a metal to dissolve in solution according to equation [1] is measured in terms of the potential produced as a result of the production of ions and elec-

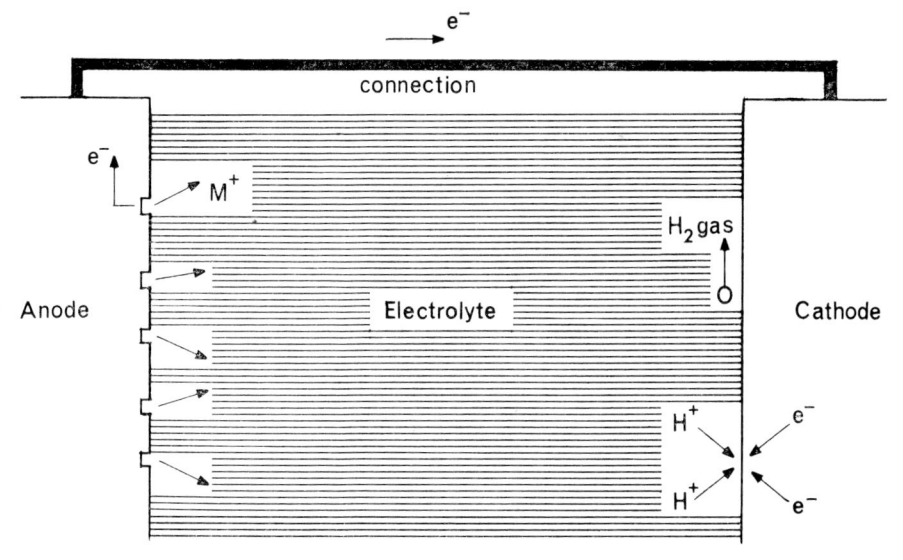

Fig.5.2. Two dissimilar metals in contact in the presence of an electrolyte. The flow of electrons from anode to cathode allows the migration of positively charged ions away from the anode, and the anode tends to dissolve. The reverse is true of the cathode where the build-up of negative charge tends to prevent solution of the cathode, and corrosion there is inhibited. Some of the charge at the cathode may contribute to reactions of the type shown.

trons. This tendency is measures in terms of the potential difference established between the metal and a standard hydrogen electrode. The potential differences measured in this way give rise to the electrochemical series shown in Table 5.1. (This series refers to measurements made using the pure metals uncontaminated with any surface films.)

The simplest kind of corrosion cell is one in which two dissimilar metals are placed in an electrolyte and are connected so that an electric current can flow between them (Fig. 5.2). Under these circumstances, one metal will act as 'anode' and the other as 'cathode'. In such a galvanic cell, the electrons flow along the connecting circuit from the anode to the cathode. At the anode, the reaction [1] is driven to the right because the electrons are removed by flowing to the cathode. This results in more metal ions going into solution in the electrolyte at the anode, and corrosion of the anode takes place. At the cathode, there will also be a tendency for the reaction [1] to take place, but the current of electrons being received by the cathode will force this reaction to the left. The tendency for the cathode to dissolve is therefore minimized, and the cathode is protected.

Several reactions can occur at the cathode of a galvanic cell. At the cathode, the electrons may react with positively charged ions in the electrolyte, and a typical reaction causes the liberation of hydrogen gas at the cathode. Thus hydrogen ions are formed in aqueous solution by the reaction,

$$H_2O \rightleftharpoons H^+ + OH^-$$

and these contribute to the reaction

$$2H^+ + 2e^- \rightleftharpoons H_2$$

at the cathode. Another very important reaction can occur at the cathode in this type of electrolyte, and this is the reaction which occurs in the rusting of iron, when iron forms the anode of a galvanic cell. In this reaction, water and oxygen combine at the cathode to produce hydroxyl ions,

$$2H_2O + O_2 + 4e^- \rightarrow 4(OH)^- \qquad [2]$$

and the Fe^{3+} ions produced at the anode react with these to form ferric hydroxide (rust):

$$Fe^{3+} + 3(OH)^- \rightarrow Fe(OH)_3.$$

With reference to the electrochemical series (Table 5.1), any metal in this series tends to be anodic with respect to metals below it. Conversely, a metal will tend to act as cathode in a corrosion cell if it is joined to a metal in a higher position in the series. The electrochemical series is an experimentally determined series which has been measured under precisely defined conditions and it refers to pure metals uncontaminated by, and unprotected by, any surface films. For practical applications a number of galvanic series have been determined which present essentially similar information for commonly used metals and alloys in a number of practical electrolytes. A galvanic series should always be consulted in considering specific corrosion problems since the galvanic series relate more closely to the practical situation.

Following from the basic electrochemical nature of corrosion, it is clear that corrosion may be encouraged if a situation is set up in which anodic and cathodic areas are produced. This does not necessarily require the contact of dissimilar metals. Different areas within the same piece of metal act as anodes and cathodes. Often as corrosion proceeds at anodic areas these areas become protected by corrosion product and tend to become less anodic with respect to adjacent cathodic areas. Gradually the situation becomes reversed, the areas that were originally cathodes becoming anodes with respect to the corroded areas. This process accounts for the relatively uniform corrosion of a pure metal. Anodic and cathodic areas are formed in a pure metal largely as a result of the polycrystalline nature of metals. The boundaries between the metal crystals (grains) will tend to be anodic and attack may initially concentrate at the grain boundaries. Similarly the differing orientation of neighbouring grains will result in slight differences between them, which will tend to result in the existence of anodic and cathodic areas.

Whilst pure metals do corrode, the electrochemical processes described above indicate that the process would be expected to be much slower than is the case when two different metals are in contact. This is because the internal differences which give rise to anodic and cathodic areas in a pure metal are far smaller than the differences in the case of dissimilar metals in contact.

Many of the metals used in practice are alloys of two or more elements, and such alloys can be either single-phase or multiphase in structure. A single-phase alloy is one in which the atoms of the different metals are combined together in a single crystal structure. In a two-phase or multiphase alloy the different phases are physically distinct. In a two-phase alloy the separate phases can form anodic and cathodic regions. This suggests that two-phase alloys would be expected to be less corrosion-resistant than pure metals, whilst homogeneous single-phase alloys may be either more or less corrosion-resistant than the constituent metals.

Another cause of differences in internal energy in any metal is the process of deformation. The differences in internal energy produced as a result of deformation will tend to produce galvanic cells, the deformed areas being anodic with respect to the undeformed areas.

Finally, the processes of corrosion are encouraged by reactions of the type given by equation [2] above, namely,

$$2H_2O + O_2 + 4e^- \rightarrow 4(OH)^-.$$

This type of reaction occurs at the cathode of the galvanic cell whilst the corrosion takes place at the area which supplies electrons, i.e. the anode. Thus the presence of oxygen and moisture at the cathode encourages corrosion of the anode (by consuming electrons). The corrosion, therefore, takes place where the oxygen concentration is least. This reaction indicates a type of corrosion problem which often occurs in a building situation, since it predicts that corrosion will take place in protected areas when surrounding areas are unprotected. Typical examples of this are corrosion of areas covered by fastening plates, bolt-heads, etc., and corrosion taking place in deep cracks and pits where the oxygen concentration is less than on the exposed areas of the metal surface (Fig. 5.3).

A consideration of the electrochemical principles indicates the situations which might give rise to corrosion problems. However, for many metals used on buildings the situation is less hazardous than indicated above. A number of metals (for example, stainless steel, alumi-

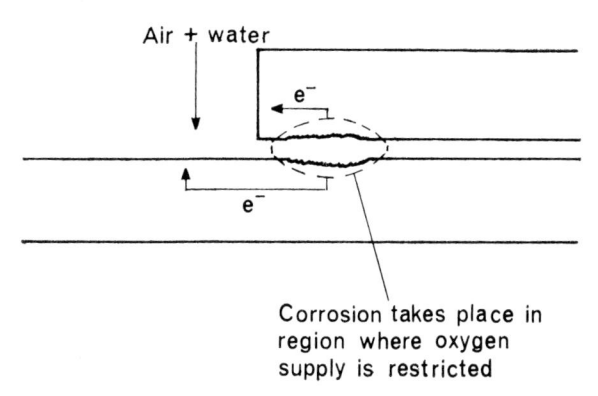

Fig.5.3. Example of corrosion taking place in regions where the oxygen concentration is low. Equation [2].

nium) become passive on exposure to the atmosphere; the metal acquires a surface film which isolates the metal and effectively protects the metal from corrosive attack. A number of metals (for example, copper, lead, zinc) which initially corrode on exposure to the atmosphere, acquire a protective layer of corrosion product, and this also serves to isolate the metal from further attack.

In the architectural use of metals externally, the fundamental consideration is the effect of atmospheric exposure on the metal. However, in any building application other potential hazards exist. It is necessary to avoid the contact of dissimilar metals unless these are known to be compatible. In all cases, therefore, the detailing of the structure must be checked for possible galvanic action. In addition, in certain circumstances, metals can undergo corrosion as a result of contact with other building materials or as a result of contaminated rainwater running off these. Normally, interaction problems are associated either with damp or uncured cements, mortars, or plasters, or with the effects of acid attack from certain types of timber particularly under moist conditions. However, most metals have gained their widespread use because they are generally compatible with other building materials.

ALUMINIUM AND ALUMINIUM ALLOYS

The aluminium industry now produces a wide range of alloys which are available in wrought and cast forms. The range of properties and uses of these alloys is equally extensive and for the purpose of this chapter the alloys will be considered as belonging to four main categories. These are the various grades of pure aluminium, the aluminium alloys that are not heat-treatable, the heat-treatable alloys, and the casting alloys.

Selection of alloys and finishes

The wide range of alloys available has resulted in aluminium being used both structurally and as a cladding metal on the external surfaces of buildings.

The super-purity grades of aluminium ($\sim$ 99·99 per cent Al) in the fully soft condition have a ductility comparable with that of lead and are used for fully supported roofing and for flashings. Aluminium shows work hardening on deformation so that the pure grades can also be used in situations where some stiffness is required, although they are unsuitable for structural applications. A major advanatage of the pure grades of aluminium is their excellent corrosion resistance.

The actual selection of alloys for structural purposes is covered by recommendations published by the Institution of Structural Engineers (1), and there are a number of relevant British Standards which give recommendations regarding the alloys suitable for particular applications.

Of the non-heat-treatable alloys, the aluminium/$1\frac{1}{4}$ per cent manganese alloy (NS3) is normally used for corrugated and troughed sheet roofing and cladding on buildings. This alloy may be used in the hard temper for such applications. Other non-heat-treatable alloys include those containing from 2 to 7 per cent magnesium, and these are available with comparatively high strength.

The heat-treatable alloys are normally used where high strength is required, and these are alloys based on the aluminium/magnesium/silicon and aluminium/copper/magnesium/silicon systems. The strength of these alloys is produced by a carefully controlled process of heat treatment. At temperatures approximately 100 °C below the melting-point, these alloys possess a single-phase structure; the alloy consists of a homogeneous solid solution of the constituent metals. Under conditions of equilibrium at room temperature, the alloys possess a two-phase structure. The primary phase is a solid solution of the constituent metals: the second phase comprises the precipitated intermetallic compounds which are dispersed throughout the alloy. The heat treatment usually applied to the heat-treatable aluminium alloys consists of heating the alloy to a temperature around 500 °C followed by rapid cooling from this temperature by quenching into water or oil. This stage is known as solution treatment since the alloy thereby becomes a single-phase solid solution. This solid solution is not stable, and on ageing at room temperature, the excess elements or compounds tend to precipitate out as a finely dispersed second phase. The process of precipitation is often accelerated by a precipitation heat treatment at temperatures around 150 °C. The finely dispersed precipitate results in the alloy being harder and stronger.

The high-strength heat-treatable alloys are used for structural purposes and for fastenings. The corrosion resistance of the alloys containing larger percentages ($\sim$4 per cent) of copper is less than that of pure aluminium. For this reason a number of the high-strength alloys used externally are clad with a layer of super-purity aluminium.

The aluminium alloys used for casting usually contain silicon as an alloying element in percentages up to $\sim$12 per cent. There are also aluminium/silicon/copper and aluminium/silicon/magnesium casting alloys.

In addition to the as-manufactured finish of the metal, a range of additional finishes may be applied to aluminium and its alloys. Among these, anodizing is the best known for architectural applications. This is a process in which the thickness of the surface film is increased by electrolytic oxidation. The film so formed is a suitable base for dyeing, and by including various organic dyes in the electrolytic bath, the aluminium may be anodized in a range of colours. The oxide layer must be sealed after anodizing. It is normally recommended that the anodic film formed in this way should be at least 0·001 in (0·025 mm) for external applications in most atmospheres.

Other surface finishes are also used with aluminium. Conversion coatings are applied in which the aluminium surface is treated with solutions of chromates, phosphates, and fluorides. Various forms of vitreous enamelling and stove enamelling are also used with aluminium. Additionally, aluminium forms an excellent base for painting provided that proper pretreatment is carried out. This normally includes degreasing, the use of an etch primer, followed by the use of a leadless primer such as zinc chromate. More recently a number of factory-applied finishes have been produced for architectural applications. These are essentially polymeric finishes which are baked on to the surface of the aluminium. These finishes have wide application in roofing and cladding.

Effect of atmospheric exposure

The corrosion resistance of aluminium and of aluminium alloys is due to the formation of a thin invisible film of oxide on the surface of the metal. This film is approximately $1·3 \times 10^{-5}$ mm ($0·512 \times 10^{-6}$ in) in thickness and is continuous over the whole surface of the metal. It forms continuously and immediately over scratches and over freshly cut metal surfaces. This oxide film effectively isolates the metal from further attack. Prolonged exposure to the atmosphere—particularly in industrial areas—results in the formation of spots of a white crystalline corrosion product on the surface of the metal, and depending upon the amount of soot in the atmosphere, the metal acquires a greyish or darker appearance.

The atmospheric corrosion of metals is constantly being assessed on a world-wide basis, and the ASTM has published the results of extensive surveys carried out by several workers on the atmospheric corrosion of metals (2, 3). In general, such tests are carried out by measuring the weight loss and/or depth of pitting on standard panels of the metals. The panels are exposed on special racks situated in a number of rural, urban, industrial, and marine environments.

The basic properties which are required of metals on building surfaces are that they should remain weatherproof in the case of cladding and roofing, and also that they should retain adequate strength. Corrosion test results are available which give an indication of any variation in tensile strength as a result of atmospheric exposure.

Tests show that the atmospheric corrosion of aluminium is greater in severe industrial environments than in the rural and marine environments. Carter (4) has reported on the results of atmospheric exposure over a six-year period at industrial, marine, and rural sites in Britain, for four aluminium alloys. McGeary *et al.* (5) have reported on the results of a parallel investigation in the USA and have compared these with the British results. The industrial sites chosen in Britain were recognized as possessing exceptionally aggressive atmospheres. The results for these severe industrial sites show that the rate of attack on the groundward-facing surfaces of panels is greater than on the surface which is washed by rainfall. It is suggested by the authors of these papers that this is due to the build-up of chemical contaminants on the surfaces unwashed by rainfall. (Condensation can occur on the groundward-facing surfaces of such exposed panels.) This attack on the groundward-facing surfaces does affect the reported corrosion rates and will make these higher than would normally be observed if the unwashed surface were protected from such a combination of condensation and contamination.

The results given in the papers of McGeary *et al.* and of Carter show that from weight-loss measurements the rates of attack at the British sites were 0.0057×10^{-3}—0.012×10^{-3} in/year (0.145×10^{-3}—0.305×10^{-3} mm/year) for rural sites, 0.007×10^{-3}—0.023×10^{-3} in/year (0.178×10^{-3}—0.584×10^{-3} mm/year) for the marine sites, and 0.085×10^{-3}–0.355×10^{-3} in/year (2.17×10^{-3}–9.02×10^{-3} mm/year) for the industrial sites. The depth of pitting on the rainwashed surfaces of the panels after six years was in the range 3.2×10^{-3}–6.6×10^{-3} in (81.2×10^{-3}–167.6×10^{-3} mm) in the industrial locations, 1.0×10^{-3}–3.2×10^{-3} in (25.4×10^{-3}–81.3×10^{-3} mm) in the marine locations, and 1.0×10^{-3}–4.4×10^{-3} in (25.4×10^{-3}–111.76×10^{-3} mm) in the rural locations. These results include both super-purity aluminium and three aluminium alloys, one of which was a clad alloy. It is reported that the super-purity grades of aluminium showed approximately half the rate of weight loss at any one site than the other alloys tested. The rate of uniform corrosion tended to decrease after intitial exposure and the depth of the pits on the skyward surfaces tended to remain fairly constant after the first two years.

Loss of tensile strength results from the corrosion of these alloys, but it is most significant (and in certain cases serious) in industrial atmospheres, and generally very small (~ 5 per cent) or insignificant in milder locations. The use of aluminium alloys structurally in exceptionally aggressive environments should be considered at the design stage in the light of the detailed data available.

If the uniform corrosion results are taken as a basis for calculation, the time for a 22 SWG (0·028 in, 0·711 mm) sheet to corrode to half its original thickness would be between approximately 40 and 2400 years depending upon the alloy used and the nature of the exposure site. Additionally, there would be some pitting which could considerably affect the useful life. The indications are that the pits do not deepen much after the first few years except in the most aggressive environments, and even in such environments there is evidence of a self-limiting effect. It would therefore seem likely that after a long building life these pits would still not perforate the metal, although pitting—which can also affect the mechanical properties—is likely to be the determining factor in controlling the useful life of the metal.

Mattsson and Lindgren (6) have reported on the weathering characteristics of a range of hard-rolled aluminium alloys exposed at a number of sites in Sweden. The results of pit depth measurements indicate that after ten years' exposure the pit depths varied from 60 to 210 μm ($2 \cdot 5 \times 10^{-3}$ to 8×10^{-3} in) depending on environment. These workers report that during these tests the rate of pit growth decreased from between 20 and 70 μm ($0 \cdot 8 \times 10^{-3}$ and $2 \cdot 8 \times 10^{-3}$ in) during the first year to between 0 and 5 μm (0 and $0 \cdot 2 \times 10^{-3}$ in) during the tenth. The self-limiting effect of the corrosion is therefore apparent in these results. On the basis of these results, pitting of a 22 SWG (0·028 in, 0·711 mm) sheet to a depth corresponding to half its thickness would take upwards of fifty years depending upon the atmosphere.

There is considerable interest in the use of clad alloys for architectural applications. Cladding is normally associated with certain heat-treatable, high-strength alloys, and Carter has reported on alloys clad with pure aluminium. A cladding layer of 3×10^{-3} in ($76 \cdot 2 \times 10^{-3}$ mm) gave sacrificial protection to the underlying metal during the six-year exposure period in all but the most severe sites. Likewise, Mattsson and Lindgren have reported that for the clad alloys they tested (these were aluminium 1·2 per cent manganese alloys clad with an aluminium/

zinc alloy) the sacrificial protection of the cladding was maintained. One particularly severe site was chosen for some of the British tests reported by Carter and here corrosive attack and loss of strength were of importance. Reference should be made to these results when considering applications of aluminium in the most severe industrial environments.

The various finishes used with aluminium generally improve the corrosion resistance. Anodized films can become pitted and damaged if allowed to remain unwashed for long periods particularly in areas where there is a precipitation of corrosive products from the atmosphere. Generally, the anodic coating becomes more durable as its thickness is increased and for this reason thicker coatings are specified in aggressive atmospheres. With coloured anodized finishes—as with other coloured finishes—it is advisable to ensure that the colours do not fade significantly on exposure to sunlight. The relevant British Standard (BS 1006) describes a method of test for organic dyes.

The various organic coatings and painted finishes protect the underlying metal completely, and the durability and general performance of the coating is the determining factor.

Performance of aluminium on building surfaces   The corrosion resistance of aluminium is seen to be excellent as regards normal atmospheric exposure. Furthermore, the corrosion-resistance of aluminium may be improved by anodizing or other protective treatments.

It is a requirement of good building practice in using aluminium that precautions are taken to prevent problems of interaction with other building materials, and that the design is such as to effectively eliminate risk of condensation occurring under roofing and behind wall cladding.

Maintenance of aluminium should be in accordance with the manufacturers' recommendations. Normally, aluminium roofing requires no maintenance. The general recommendation for window framing or curtain walling is that the metalwork should be washed at fairly regular intervals. This helps to remove any foreign matter, and also effectively prevents damage occurring to anodized finishes. Such washing is often criticized on economic grounds, and the generally low rates of corrosion might suggest that it could be avoided. However, areas of curtain walling and window frames do not get effectively washed by the rainfall, and failure to wash the surface will ultimately result in the breakdown of the anodized film with a general deterioration in appearance.

Other finishes will also benefit from regular washing to remove dust, grime, and soot. Such cleaning should be carried out in accordance with the precise instructions from the manufacturers, and selection of the right cleaning aids can be of importance.

The earliest application of aluminium on buildings was for the domes of the cupolas and semi-cupolas of the Church of San Gioacchino, Rome. This was constructed in 1897 and an examination in 1966 quoted by the Aluminium Federation (7) indicates that this roof is still entirely satisfactory and that corrosion is superficial. There are a number of other early applications of aluminium which are still in good condition and which emphasize the good corrosion resistance of aluminium used externally.

The super-purity grades of aluminium are used on fully supported pitched and flat roofing and also for flashings. The application of aluminium in this way is covered by the relevant Code of Practice (CP 143). Such roofing may be applied following the traditional practice using standing seams or batten rolls at joints. There are also systems based on the traditional practice which are available from specialist roofing manufacturers, and these systems include both untreated aluminium and aluminium coated with a number of plastic finishes.

Corrugated and troughed forms of aluminium used for

roofing and cladding are available from the manufacturers in various profiles and finishes. The relevant British Standards (BS 2855 and BS 3428) deal with this type of sheeting, and manufacturers produce a range of profiles and fixing systems. The Code of Practice CP 143 describes the use of this type of sheet.

Provided the requirements of good practice are followed, aluminium roofing of all types should possess a very long life. When aluminium is used on nominally flat roofing—usually following the traditional practices for fully supported metal roofing—the pitch should be sufficient to prevent water lying on the roof. Pools of water will tend to encourage a concentration type of corrosion cell and may give trouble particularly in areas where the rainfall is strongly acidic. Otherwise, the washing action of rainwater is beneficial in removing the small particles of soot and grit.

Condensation on the underside of roofs should be avoided and this is accomplished by the various systems of vapour barriers and ventilation that are consistent with good practice. Condensation on the underside of single-skin roofs associated with industrial situations—as for example, in chemical plant—may give rise to problems of corrosion. These must be seen as exceptional situations and reference should be made to the results reported (2, 3) by the ASTM.

Fixings for aluminium building components should ideally always be of aluminium or aluminium alloy so as to avoid any galvanic corrosion hazard. Otherwise, as a general guide, the non-magnetic stainless steels are generally compatible with aluminium although contact may not be desirable in marine or severe industrial environments. The practical galvanic series indicate that zinc tends to be slightly anodic with respect to aluminium and this implies that aluminium may tend to accelerate corrosion of zinc. Generally, however, galvanized (zinc-plated) fixings can be suitable for use with aluminium and are allowed by CP 143. It is desirable to consider the likely life of such fixings when they are used in exposed aggressive situations, since when the protective plating is consumed further trouble may occur when the underlying metal is exposed.

Aluminium tends to be anodic with respect to most other metals commonly used on buildings, and care is necessary to avoid contact with those metals known to significantly accelerate the corrosion of aluminium. Direct contact between aluminium and steel should preferably be avoided. This is particularly important in situations where some condensed moisture may accumulate or where there is excessive industrial or marine pollution. When unplated steel fixings need to be used with aluminium, a system of insulating washers should be used to isolate the metals from direct contact. Direct contact between aluminium and steelwork may be avoided by the use of a bituminous or plastic material between the surfaces or by various painting procedures. The aluminium/steel combination has not proved a particularly serious hazard, but the simple precautions to avoid trouble are well worth while.

In the case of aluminium, by far the most hazardous form of bimetallic corrosion occurs as a result of interaction with copper; the copper accelerating corrosion of the aluminium. Similarly, the run-off from copper roofs or water discharged through copper pipes should not be allowed to wash over or splash against aluminium. Notable failures have occurred to aluminium rainwater goods carrying the water from copper roofs, and the serious attack of aluminium pipes carrying copper-bearing waste water has also been observed. These hazards apart, aluminium rainwater goods give entirely satisfactory service.

The compatibility of aluminium with most of the concretes, mortars, and plasters used in buildings is generally satisfactory, and any corrosion effects are usually restricted to the action of uncured cements or plasters or to very damp situations. The results of a very detailed study of the effect of embedding aluminium alloys in building materials have been reported by Everett (8) and by Jones and Tarleton (9). Corrosion of the aluminium can result in the cracking of the embedding medium in certain circumstances, and this effect could be undesirable in a practical situation. The corrosion resistance of the aluminium depends upon the alloy used and the type of cement or plaster product adjacent to it. The normal recommendation is that the aluminium be protected with two coats of a bituminous paint at the area of contact. This precaution is, of course, particularly advisable in damp situations. Similarly, it is advisable to avoid contact between aluminium and damp, porous brickwork or stonework, preferably by the use of spacers or an impermeable membrane, or by the use of bituminous paint.

As with most metals, there is no interaction between aluminium and dry, seasoned timber. But, as with other metals, precautions are advisable when using unseasoned, acid timbers or in situations where the wood is damp. Oak and western red cedar show an acid reaction when unseasoned, and corrosion of aluminium roofing or flashings can result from water draining off unweathered shingles. In such situations, the aluminium should be protected with bituminous paint. In damp situations, joints between timber and aluminium are well protected by painting the timber with an aluminium paint, or, in aggressive environments, by interposing an impermeable gasket membrane between the aluminium and the wood.

Aluminium is generally compatible with most wood preservatives and the growing use of timber preservative treatments should not lead to any problems in the application of aluminium, provided that the preservative specified has been selected as being satisfactory. Certain preservatives containing water-soluble copper or mercury salts can have an aggressive reaction to aluminium. Preferably a compatible preservative should be specified, or else the aluminium insulated from the timber by plastic, bituminous paint, etc.

ZINC

Zinc is used architecturally in the form of rolled sheet and strip, and also as the basis of a number of protective coatings. The use of zinc sheet and strip for roofing is not so widespread in Britain as in many European countries, but zinc is widely used as a protective coating for steelwork.

Zinc is anodic with respect to iron and consequently provides some sacrificial protection when plated on to a steel surface. When the underlying steel surface is exposed at scratches the zinc corrodes preferentially and the steel remains free from attack. There are a number of ways in which zinc is applied to metal surfaces. These include hot-dip galvanizing, electrolytic processes, metal spraying, sheradizing, and the use of zinc-rich paints. Hot-dip galvanizing, in whcih the component—after proper surface cleaning—is dipped in a bath of molten zinc, gives a relatively thick surface coating of zinc (0·003–0·005 in, 0·076–0·127 mm). A further advantage of this method is the excellent surface coverage which the zinc gives to the underlying steel. The zinc bonds strongly to the underlying metal and a layer of zinc/iron alloy is formed between the zinc surface layer and the base steel. Of the other processes of applying a protective zinc coating the electrolytic process is usually applied when a thin surface covering (0·0001–0·001 in, 0·00254–0·0254 mm) is required. The metal-spraying processes give thicker coverings in the range 0·004–0·02 in (0·102–0·508 mm) in thickness. These are rather more porous but tend to become impermeable with time as any porosity becomes filled with corrosion product. Sheradizing is usually applied to small components by heating these in a drum of zinc dust at a temperature just below the melting-point of zinc. A zinc diffusion coating is formed on the articles and this gives a continuous uniform coating between 0·0005 and 0·0015 in (0·0127 and 0·0381 mm) in thickness. The thickness of the coating is closely controlled. The zinc-rich paints are based on zinc dust and contain a relatively high percentage of metallic zinc. Generally the painted coatings are approximately 0·0015 in (0·038 mm) in thickness.

Zinc sheet and strip are traditionally manufactured from commercially pure zinc, but more recently zinc/copper/titanium alloys have been developed which show superior mechanical properties to the unalloyed zinc. These alloys contain approximately 1 per cent of copper and 0·1 per cent of titanium. Notably they show a better creep-resistance and somewhat greater strength and hardness coupled with a greater ductility.

Effect of atmospheric exposure

Atmospheric exposure results in zinc acquiring a surface film of basic zinc carbonate, which protects the underlying metal and checks further corrosion. In unpolluted atmospheres zinc has a long, maintenance-free life. In industrial environments, and wherever sulphurous pollution is present, the corrosion rate of zinc increases markedly. The effect of such pollution is to form sulphurous and sulphuric acids which in turn react with the basic zinc carbonate film and convert this to zinc sulphate, which is soluble and washes off with rainfall. The zinc carbonate layer is rapidly renewed and remains protective until this is subject to further attack. The rate of corrosion of zinc is therefore dependent on the degree of acid pollution in the atmosphere.

Anderson (10) has reported in detail on the results of ten-year and twenty-year atmospheric exposure tests of rolled zinc. These tests were carried out in the USA. It was concluded that the corrosion of zinc was lowest in dry, pure atmospheres and highest in industrial atmospheres. The average penetrations—from weight loss measurements—varied from 0.007×10^{-3} in/year (0.178×10^{-3} mm/year) in an arid situation to 0.252×10^{-3} in/year (6.40×10^{-3} mm/year) in an industrial situation. The values for rural and coastal situations were given as 0.042×10^{-3} in/year (1.067×10^{-3} mm/year) and 0.058×10^{-3} in/year (1.473×10^{-3} mm/year). The results indicated that the corrosion rate of zinc in the atmosphere does not change with time of exposure provided that the environment does not change. Dunbar (11) in some more recent work in the USA noted slight changes in corrosion rate which were ascribed to changes in environment due to increased urbanization and industrialization around certain of the exposure sites. The results reported by Dunbar show similar corrosion rates to those reported in the earlier tests. Dunbar noted some slight roughening of the panels exposed at 30° to the horizontal, whilst some zinc/1 per cent copper alloys included in these tests showed some distinct pitting. No results have been reported in this series of tests on the zinc/copper/titanium alloys. The pit depths (or depth of roughening in the case of pure zinc) lay in the approximate range 1×10^{-3}–5.69×10^{-3} in (25.4×10^{-3}–144.5×10^{-3} mm), but the author noted the pit depth to total penetration ratio decreased with increased over-all penetration so that complete penetration of a panel would be as likely to occur by general corrosion as by pitting.

On the basis of the figures reported by Anderson the time for a 0.03 in (21 SWG, 14 ZG, 0.8 mm) zinc sheet to be reduced to half its thickness would be in the range 60–350 years in industrial, urban, and coastal situations. (The results for the arid situation would give a life almost ten times this maximum, but that is clearly a special situation.)

Performance of zinc on building surfaces

Whether the newer zinc/copper/titanium rolling alloys or commercially pure zinc are used as sheet or strip the general resistance to atmospheric attack will be similar in any given environment, and because the zinc tends to corrode at a uniform rate the life of the zinc is approximately proportional to the metal thickness. Zinc sheeting is normally available in thicknesses varying from 0·02 to 0·04 in (0·4 to 1 mm, ZG10 to 16). Generally, 0·04 in (1 mm, 19 SWG, ZG 16) is the limit of thickness because of fabrication difficulties and 0·03 in (0·8 mm, 21 SWG, 14 ZG) is widely used in roofing and cladding work.

Special surface finishes are not normally applied to zinc although, if required, zinc can be painted provided that appropriate primers are used.

Zinc is used in fully supported roofing using the traditional methods of construction, and is also applied using specialist systems based on these. As with all such forms of construction attention has to be paid to the detailing to give efficient weatherproof fixing and to allow for thermal expansion effects. In a similar way zinc is used for vertical cladding and also for weatherings and flashings around parapets.

The principal hazard to zinc—which is avoided if good practice is observed—is a form of corrosion known as white rusting. This occurs on zinc surfaces which are kept wet and which are not exposed. The corrosion product is a voluminous white powder which fails to give protection to the underlying metal. Typically, white rusting can occur on zinc sheeting and other zinc products which are closely stacked and unprotected from rainfall. Under these circumstances the moisture trapped between the sheets or other zinc products gives rise to white rusting. This can cause perforation of zinc sheeting in a matter of weeks. It is, therefore, essential to store zinc products in such a way that there is a free circulation of air around all the surfaces, and, if possible, the products should be protected from rainfall. Sheets should be separated in storage.

In all external applications of zinc sheet and strip it is essential to avoid the build-up of condensation on the underside or inner face of the zinc sheeting. In very confined conditions this could lead to white rusting followed by rapid failure, but in all cases condensation will cause some corrosive attack. Ventilation of roof spaces and the use of suitable underlays (BS 747) and vapour barriers will prevent trouble.

Because zinc is attacked more rapidly in polluted industrial atmospheres its use in these areas should, if possible, be restricted to vertical cladding and to pitched roofs. Its use on flatter roofs, from which the acid rainwater will drain off more slowly, will be less satisfactory, and increasing the slope of a nominally flat roof to a modest pitch can double the life of the zinc. Such a precaution is advantageous in any atmosphere, since the corrosion of zinc only tends to take place at appreciable rates when the zinc surface is wet. Rapid draining of a surface and consequently faster drying assist considerably in minimizing any corrosion problems.

When zinc roofs are used on buildings the effect of localized pollution should be considered. For example, vent pipes from oil-burning appliances can have an adverse effect on a surrounding or nearby zinc roof due to increased sulphurous pollution.

In the various methods of fixing zinc sheeting and cladding it is normal to use zinc and zinc-plated (galvanized) fixings. These do not give rise to any bimetallic corrosion. It is important to avoid contact between zinc and copper, and copper fixings must not be used with zinc. Similarly the run-off from copper roofs and out of copper pipes should not be allowed to pass over zinc roofs or gutters. Contact between zinc and aluminium is generally satisfactory, as is contact between zinc and lead.

In common with other non-ferrous metals, zinc shows good compatibility with other building materials. Contact with clean mortars and cements causes very slight surface attack during setting but thereafter ceases. Particularly in older buildings the presence of sulphates and chlorides in the mortars can cause attack under damp conditions. In such cases, the application of two coats of bituminous paint to the zinc should eliminate trouble. At flashings, etc., the detailing should prevent the formation of confined pockets in which moisture can accumulate and possibly cause white rusting. This effect is avoided if normal good practice is observed. Where zinc is likely to be wet for prolonged periods, as, for example, on flashings under the edge of tiles, coating with bituminous paint is advisable.

Zinc does not show any adverse reaction with most seasoned timbers, but contact with acid woods—particularly if these are damp or unseasoned—can cause trouble. Water draining from cedar shingles or from oak should not be allowed to fall on to zinc flashings, of if this is unavoidable the zinc should be fully protected with bituminous paint. There is generally no adverse reaction between zinc and the widely used timber preservatives, but preservatives which are not water soluble and which are inert to zinc are preferred. Processes involving treatment with zinc chloride are unsuitable for use with timber which is in contact with zinc.

COPPER

Copper was one of the first metals to be used by man. It occurs in a number of mineral ores and also in the metallic state in certain places.

Copper is widely used as an alloying element in both nonferrous and ferrous alloys. Architecturally it is used as commercially pure copper and is also present as a principal alloying element in the bronzes and brasses. Steels containing small percentages ($\sim$0·5 per cent) of copper are also gaining importance for architectural applications due to their enhanced corrosion resistance.

Copper in the form of sheet and strip is used for roofing and cladding of buildings following traditional practices and the specialist roofing manufacturers' schemes based on these. It is similarly used for the covering of dormers and for the valley gutters associated with copper roofs and claddings.

Effect of atmospheric exposure

A characteristic weathering phenomenon associated with copper used externally is the formation of a green patina on the surface of the metal. Most architectural schemes which use copper are designed on the basis of the copper eventually acquiring this patina. The determination of the actual composition of this patina has been the subject of considerable study, as have methods of producing this patina artificially. The patina consists of copper hydroxide salts of the sulphate and chloride variants. Small amounts of the copper hydroxide salts of the nitrate and carbonate variants may also be present in certain circumstances. After about seventy years the surface layer becomes mineralized, the patina consisting of the copper hydroxide sulphate $Cu_4(SO_4)(OH)_6$ variant which is identical with the mineral brochantite and possibly the copper hydroxide chloride variant $Cu_2Cl(OH)_3$ which is identical with the mineral atacamite. The sulphate variant is generally predominant, but the chloride variant is present on copper weathered in marine locations.

Initially on exposure to the atmosphere copper acquires a dark brown colouring which becomes even over the whole surface after about six months. This is the initial oxide film forming over the copper surface, and this may become very dark in urban areas. Thereafter, after several years' exposure, the green patina begins to form, and the whole area of metal acquires an even or a mottled green colour. The normal time for patina formation is five to ten years depending on the degree of pollution, the moisture present in the air, and the temperature. The formation of the patina is an electrochemical process like other corrosion mechanisms, and therefore proceeds in the presence of a film of moisture on the surface of the metal. The formation of the patina also requires some atmospheric pollution, either from industrial and urban sources (giving sulphurous pollution) or from marine sources (giving chloride pollution). It is, in fact, only in the most remote areas of the world where there is not sufficient contamination of the atmosphere to cause patina formation. The reaction leading to patina formation also proceeds more rapidly at higher temperatures. For these reasons in cold atmospheres in which there is virtually no atmospheric pollution—that is in regions remote from industrial areas, from urban areas and from the sea—the patina may not appear within a reasonable time and in fact may not appear within the life of the building. Such conditions are, of course, exceptionally rare, but in certain regions in the Swiss Alps and in remote parts of Sweden isolated copper roofs have shown no sign of patina formation after several decades.

The actual rates of corrosion of copper and copper alloys have been reported by Tracy (12) and by Thomson, Tracy, and Freeman (13) based on twenty-year atmospheric exposure tests in the USA. In an extensive twenty-year investigation of the rates of corrosion of eleven copper alloys, Tracy concluded that the corrosion rates were highest in industrial atmospheres (averaging less than 0.1×10^{-3} in/year (2.54×10^{-3} mm/year)) and least in rural atmospheres (averaging less than 0.025×10^{-3} in/year (0.635×10^{-3} mm/year)). Thomson, Tracy, and Freeman examined eleven brands of copper at four exposure sites in Connecticut and concluded that the industrial atmospheres were more corrosive than the marine which in turn were more corrosive than the rural atmospheres. The mean corrosion rates were of the order of 0.050×10^{-3} in/year (1.27×10^{-3} mm/year) over the twenty-year period. Although very slight variations in rates of corrosion with different brands and purities of copper were found these were not regarded as of practical significance. These results reported by Thomson, Tracy, and Freeman were obtained from loss of weight data and from loss of strength data on sheet metal and from loss in strength data and increase in electrical resistance data in wires. For sheet metal the reported mean corrosion rates were around 0.035×10^{-3} in/year (0.889×10^{-3} mm/year). Thomson (14) has recently reported on the corrosion rates of eighteen copper alloys exposed at four test sites in the USA. The mean corrosion rate from weight loss data was on the order of 0.05×10^{-3} in/year (1.27×10^{-3} mm/year). Mattsson and Holm (15) have reported on an extensive investigation of the corrosion rates of copper alloys in seven-year exposure tests in Sweden. The average penetration calculated from weight losses varied between $0.2 \times 10^{-3} - 0.6 \times 10^{-3}$ mm/year ($0.0079 \times 10^{-3} - 0.0236 \times 10^{-3}$ in/year) in rural atmospheres to $0.6 \times 10^{-3} - 1.1 \times 10^{-3}$ mm/year ($0.0236 \times 10^{-3} - 0.0433 \times 10^{-3}$ in/year) in marine atmospheres to $0.9 \times 10^{-3} - 2.2 \times 10^{-3}$ mm/year ($0.0354 \times 10^{-3} - 0.0866 \times 10^{-3}$ in/year) in industrial atmospheres.

Applications of copper on buildings

Copper is used for roofings, claddings, weatherings, and flashings on the external surfaces of buildings following the developments of several centuries of traditional practice. A range of copper alloys, including brasses and bronzes, are used externally on door-frames, balustrades and window surrounds, and other fittings.

The use of copper sheet and strip for fully supported roofing following traditional practice requires the copper to be in the fully annealed (soft) condition. With this method of construction, allowance is made for thermal movement at joints with maximum lengths of copper of approximately 7 ft (2·1m) between joints. More recently, constructional methods have been developed which use the copper in longer length strips, and these use the copper in the $\frac{1}{8}$-hard to $\frac{1}{4}$-hard temper. The metal is therefore slightly more rigid, and this rigidity coupled with special fixing techniques allows thermal movement to be transmitted over longer lengths and strips up to 26 ft (7·9m) long may be laid.

There are two possibilities as regards 'natural' finishes for copper. The first of these is the 'patina' finish and this normally appears as a result of the processes of weathering. However, a number of attempts have been made to produce artificially patinated sheeting, and it is now claimed that these are satisfactory. The advantages of artificial patination for a new building are that the several years of exposure to acquire the green finish are unnecessary, and the building should have its designed appearance from the time of completion. The advantages are more significant when repairing existing copper roofs in that a better match should be obtained by using pre-patinated sheet.

The second type of finish is the use of an inhibited protective lacquer to preserve the polished finish of the copper. Such a lacquer must contain an inhibitor otherwise corrosion will proceed underneath the film around slight scratches and imperfections in the coating (equation [2]). These inhibited lacquers have been in use for several years and are successful in preserving a polished copper finish. These lacquers will need renewal at intervals, but the life will depend on the conditions of exposure.

Constructional details and precautions	The general precautions necessary in the use of copper are as for other metal roofing and cladding. The copper sheet and strip used in roofing are laid on suitable underlays as prescribed in BS 747 and CP 143. As with other roofing, condensation should be avoided by allowing

Metals

slight ventilation of roof spaces and other good constructional practice. Galvanic attack to copper is avoided by the use of copper nails or brass screws and fixings. Copper shows good compatibility with other building materials, and contact between copper and clean cements and mortars is normally quite satisfactory; copper is used satisfactorily as a dpc (BS 743).

Copper is compatible with building timbers, but precautions may be advisable in the case of rainwater discharging from cedar shingles on to copper. The simple precautions of using bituminous paint on the copper where the water discharges from the shingles should be satisfactory in these exceptional situations.

Copper is a relatively noble metal and the corrosion of other metals in contact with copper will tend to be increased. for this reason copper must not be used in conjunction with aluminium or zinc as it will cause corrosion of these metals. The rainfall running over copper must not be allowed to wash over aluminium or zinc rainwater goods. Similarly, contact between galvanized steel or steel fixings and copper would not be satisfactory and should be avoided.

Copper roofing and cladding should have the appropriate copper rainwater goods designed in conjunction with the general scheme. Staining can arise from water discharging from copper over other building materials, but a well-designed rainwater collection system will obviate this problem.

Performance of copper on buildings

Copper has traditionally been used for the roofing of large buildings intended to last more than a century, and its performance in these applications has been entirely satisfactory. The corrosion results reported in the literature suggest that the rate of corrosion of copper is rarely greater than 0.050×10^{-3} in/year (1.27×10^{-3} mm/year) which suggests that a copper roof of 0.028 in (22 SWG,

0·711 mm) thickness would be corroded to one-half of its original thickness after approximately 280 years. These figures do not take into account the fact that the surface of copper tends to become fully mineralized after approximately seventy years and thereafter corrosion should effectively cease. All these figures suggest that a copper roof is likely to have an exceptional life, and failure due to straightforward atmospheric corrosion is unlikely within the life of the building. This is consistent with centuries of experience of copper roofing.

In addition to the outstanding durability of copper in the atmosphere the metal also possesses exceptional fire-resistance and serious fires have been contained under copper roofing.

LEAD

Historically one of the earliest applications of lead was for the flooring of the Hanging Gardens of Babylon, and it has long been used as a metal for high-quality roofing and cladding of building surfaces.

Lead sheet and strip used for building surfaces are manufactured in Britain to BS 1178 and are manufactured from lead of 99·9 per cent purity. In the USA antimonial lead containing 6–7 per cent antimony is more commonly used. The antimony additions give increased stiffness and strength to the alloy. Lead sheet and strip are manufactured either by traditional rolling or by more modern continuous casting processes.

Because of its excellent corrosion resistance lead is used for coating iron and steel. Lead-coated copper is also used. All the normal methods of coating are used but lead alloys containing $\sim$ 2 per cent tin are used for the hot-dipping processes. Sheet lead cladding of steel, for example in glazing systems, is also used to protect the underlying metal.

Metals

Lead and lead coatings form an excellent base for lead-based paints, although aluminium-based paint systems should not be used.

Effect of atmospheric exposure

When freshly cut, lead has a shiny metallic appearance which rapidly dulls due to the formation of an oxide layer on the surface of the metal. Prolonged exposure results in the formation of protective layers of lead carbonates and sulphates on the surface. This gives the lead a whitish-grey appearance in clean atmospheres. It may darken in industrial areas. Lead is notable for its corrosion resistance in all atmospheres, and the unprotected metal shows excellent corrosion resistance in industrial and marine environments as well as in less-aggressive situations.

Hiers and Minarcik (16) have reported on the twenty-year corrosion rates of lead exposed at seven sites in the USA. From their weight-loss measurements over twenty years there was an indication that the rates of corrosion were falling in all but the sea coast situations. Both lead and antimonial lead (1 per cent antimony) were extremely durable with corrosion rates around 0.02×10^{-3} in/year (0.508×10^{-3} mm/year). At this rate No. 5 lead sheet (2·24 mm, $\frac{3}{32}$ in) would corrode to one-half its thickness in 2500 years.

Performance on building surfaces

The lead used in Britain is normally of the 99·9 per cent composition, and traditionally it is applied over tongue and groove boarding with a suitable building paper (BS 1521) or felt (BS 747) underlay. Such lead sheet is very soft and ductile, and on flat and pitched roofs it tends to settle into close contact with the supporting boarding. Lead sheet is also used for cladding around parapets and on vertical surfaces generally, and its use in these applications is tending to increase. It is possible to fix the lead sheet on vertical surfaces with standing seams to emphasize the vertical lines and good examples of this practice exist. This is possible using 99·9 per cent lead as is normal practice in Britain, although in the USA the antimonial leads, which are rather more rigid, are widely used. There are now specialist systems using lead sheeting for roofing and cladding which are available through the appropriate manufacturers. Lead is normally supplied in a smooth rolled finish but a number of different embossed finishes can be supplied.

Lead roofs are associated with large buildings designed to last at least a century, and lead has been successfully and widely used in such applications. Additionally, lead has been used on the roofing and flashings of many smaller buildings, and most domestic buildings incorporate some lead on the external surface. Lead is very widely used for flashings, valley gutters, flat roofs on bays and dormers, and in all these situations is known to perform satisfactorily over very long periods of time. Lead is entirely satisfactory in polluted industrial atmospheres as well as in milder situations.

Lead has useful acoustic properties and it may be used for reducing the transmission of sound. It is, therefore, useful for cladding buildings in which it is desired to either reduce exterior sounds or to contain the sounds generated within the building.

Constructional details and precautions

Lead is applied in accordance with traditional practice following the normal precautions necessary for the use of metals for cladding and roofing. Suitable underlays should be provided and it is particularly important in the use of a soft metal that no sharp projections exist on the supporting substrate. Thermal movement must be allowed for at all joints and throughout the roof design. Loose-lock seam smust be provided at joints and except in exceptional circumstances all the fixings should be concealed under the lead. Copper clips, brass screws and washers provide the normal fixing method, and any brass screws passing through lead sheeting from the outer surface are normally covered with caps of lead which are 'lead-burned' on to the surrounding lead sheet. The high density of lead, and its general lack of rigidity, requires a strong supporting structure and firm, well-designed fixings.

The problem of galvanic action is not normally significant in the case of lead. In certain aggressive atmospheres contact between lead and aluminium may be undesirable, the lead tending to encourage corrosion of the aluminium, so that contact between the two metals is best avoided.

Problems have been noted of the corrosion of lead in contact with damp, uncured cements and mortars and the usual precaution of two coats of bituminous paint will prevent such troubles. Similarly, it is recommended that lead used for dpc should be suitably protected. Lead does not react with clean cements and mortars when the curing of these is complete. The successful use of lead in flashings on brickwork proves that the problems of interaction are not usually significant provided that normal good workmanship is observed.

Contact between lead and damp acid timbers can cause corrosive attack on the lead, and lead flashings used in buildings in conjunction with unweathered cedar shingles should be treated with bituminous paint until the weathering of the shingles has been completed. Lead can also suffer acid attack from water draining off roofs overgrown with mosses or lichens. Most of these effects are relatively minor and are not likely to result in rapid failure. The practice of good construction to ensure that the water drains away efficiently is of greater importance in such situations.

A major advantage of lead is its non-staining character. The water draining from lead work will not of itself stain the surrounding structure, but it is in any case desirable to consider the possibility of incorporating gutters to claddings and other lead surface finishes to prevent streaks of rainwater forming down the surface of the building.

PLAIN CARBON AND LOW-ALLOY STEELS

Steels are used in all types and in all aspects of buildings, and it is outside the scope of this chapter to consider in detail the range of types of steel that are available. From the general aspect of weathering, the steels which are used externally on buildings fall into two main categories: namely, those which are to be protected by some type of surface treatment, and those alloy steels which are designed to require no maintenance.

The steel components present on the external surfaces of buildings range from small fixings to cladding units to which surface finishes may have been applied. Steel is also present in various structural capacities and as balustrades, stairways, and balconies.

Effect of atmospheric exposure

The rusting of steel is well known and it is generally desired to avoid this form of deterioration. There is a type of alloy steel based on the copper-bearing steels which has essentially slow rusting characteristics. These steels are gaining application in building, the colours of the corroded surface making an aesthetically satisfactory contribution towards the appearance of the building.

Recently, Thomas and Alderson (17) have reported on the corrosion rates of mild steel exposed at a number of locations in California, and have concluded that the most severe conditions as regards the corrosion of steel are found at the coastal/marine locations. The weight loss data (gathered over a five-year exposure period) show, with the exclusion of an exceptional marine site, corrosion rates in the range $0.32 \times 10^{-3} - 1.89 \times 10^{-3}$ in/year ($8.128 \times 10^{-3} - 48.0 \times 10^{-3}$ mm/year). Briggs (18) has reported the results of detailed investigations extending over twelve years at marine and industrial sites in the USA of the corrosion rates of some cast and wrought carbon and low-alloy steels. Briggs found that the corrosion rates decreased markedly with time, and that the low-alloy steels containing nickel, chromium, or copper had a superior corrosion resistance.

The architecturally used, slow-rusting, alloy steels acquire a dense coherent layer of corrosion product on the surface, and this surface layer tends to stifle further corrosive attack. It is usually claimed that these steels should not show more than $\sim 1 \times 10^{-3} - 3 \times 10^{-3}$ in ($25.4 \times 10^{-3} - 76.2 \times 10^{-3}$ mm) of penetration during the first five to ten years.

General properties and finishes

The finishes which are commonly used on steels used externally on buildings include the traditional painted finishes, the factory-applied finishes based on various organic polymers, and the metal-plated finishes. Of these latter the galvanized or zinc-coated finishes are widely used, although other metal coatings are of importance in special applications (for example, lead coverings).

The galvanizing or zinc plating of steel has had a long application in corrugated steel roofing and in applications for certain types of rainwater goods. The success of these coatings derives both from their good adherence to, and total surface coverage of, the underlying steel and also from the galvanic protection that the zinc gives to the iron. At any break in the zinc coating the zinc will still corrode preferentially to the steel (zinc being anodic with respect to iron). The use of galvanizing, or of zinc coating of steel by other methods, has grown with the need to avoid unsightly rust staining of the surrounding building structure as well as to control other serious aspects of corrosion. Galvanizing is also being used more widely as the first stage in the surface protection of steel, the galvanized surface being subsequently painted or otherwise treated.

The performance of the widely used zinc coatings can be approximately assessed from the reported corrosion rates of zinc. The maximum rate of corrosion of zinc reported by Anderson was 0.309×10^{-3} in/year (7.85×10^{-3} mm/year) for a severe industrial site in the USA. Gilbert (19) has summarized the corrosion rates of zinc and of zinc-coated steel measured in a scheme of work carried out by the British Non-Ferrous Metals Research Association. The rate of attack of galvanized steel specimens exposed in a severe industrial atmosphere was approximately 0.2 oz/ft^2 per year, and it was concluded that the life of galvanized coatings is proportional to the thickness of the coatings. A zinc coating of 2 oz/ft^2 is approximately 0.0035 in (0.0889 mm) in thickness, and the results cited above might indicate that such a coating should be consumed after approximately ten years in a severe industrial atmosphere. The reported corrosion rates for zinc show that galvanized coatings may be expected to have very much longer lives in rural, urban, and marine environments. In the milder environments, a coating of 2 oz/ft^2 may be expected to have a life of upwards of twenty years or longer.

The plastic-coated finishes applied to zinc-coated steel afford good protection to the underlying metal, and the combination of a polymer coating and galvanized protection should give a satisfactory performance. Usually for external applications the steelwork is hot-dip galvanized to BS 2989 prior to application of the final coating. In certain circumstances the steel may be zinc coated by electrodeposition. This gives a thinner coating of zinc, and greater reliance is placed on the protection afforded by the finish.

Galvanized finishes may also be painted at any time. The appropriate primers need to be used to ensure good adherence of the paint finish. A correctly applied paint system will last longer on zinc-coated steel than on uncoated steel, and problems of rusting should not arise.

The low-alloy steels developed from the copper-bearing steels are becoming more widely used for architectural purposes and do not require any finish to be applied to the exposed surfaces. These slow-rusting steels have interesting weathering characteristics, and are stronger than the plain carbon structural steels. The corrosion of these steels is essentially self-stifling, but it is usual in the detailing of buildings constructed using these steels to avoid problems of staining by ensuring that the rainwater run-off from these steels does not pass over vulnerable materials. These steels are normally protected on the inner faces of the metal and on areas in contact with other building materials by painting.

STAINLESS STEELS

The use of stainless steel on the external surfaces of buildings is now well established. The great value of stainless steel in these applications is its high strength coupled with an outstanding resistance to atmospheric corrosion. Whilst most metals show essentially good resistance to corrosion in the atmosphere, a suitable stainless steel will retain its original lustre subject to washing off of superficial grime.

General properties and and uses of stainless steel

The stainless steels are those highly alloyed steels which contain relatively large (17–20 per cent) percentages of chromium. The stainless steels used in building are those containing 17 per cent chromium, those containing 18 per cent chromium and 8 per cent nickel (18/8) and those containing 18 per cent chromium, 10 per cent nickel, 3 per cent molybdenum (18/10/3). The 17 per cent chromium alloy is a magnetic, ferritic stainless steel, whilst the 18/8 and 18/10/3 alloys are non-magnetic, austenitic stainless steels.

As a general guide, the 17 per cent chromium steel is suitable for the milder and protected situations. The 18/8 steel is suitable indoors and also outdoors where appearance is unimportant. For aggressive situations and for all external applications where it is required that the components maintain their initial appearance, it is advisable to use the 18/10/3 steels. These 18/10/3 steels have exceptional durability in the most aggressive atmospheres.

Stainless steel is used externally both as a non-load-bearing cladding metal and also in load-bearing sections for structural purposes. When stainless steel is specified as a cladding metal it is used in thin section to protect the underlying structural material. Both methods of using stainless steel have been applied in curtain walling systems.

The stainless steels have been applied to the more complex sections associated with window framing, curtain walling, shop fronts, etc., where the corrosion resistance of these alloys makes them particularly suited to these complex constructions. Stainless steels are used for cladding. Additionally, a very important use of stainless steels is for fastenings and fixings, which have a wide application being compatible with certain other metals.

Effect of atmospheric exposure

The stainless steels owe their exceptional corrosion resistance to the formation of a thin oxide film which is always present on the surface of the stainless steel, and which is self-repairing when damaged. Whilst all the stainless steels are very corrosion-resistant for architectural purposes, the 18/10/3 grade is normally recommended for external applications in the aggressive atmosphere of Britain. This steel would not be expected to show any surface deterioration. The 18/8 steels might possibly suffer some slight rusting and superficial deterioration in exceptionally aggressive atmospheres, particularly if the surface of the metal were to become covered with a layer of industrial or marine contaminants. The corrosion is only likely to be superficial, and it would seem to be possible to prevent this kind of attack by regular washing. The 18/8 steels are more widely used in countries outside Britain. If the 18/10/3 steel is used, regular washing is desirable chiefly to maintain a lustrous appearance. Even if such a steel is left for several years without cleaning there should be no deterioration of the metal.

Finishes and constructional precautions

There are several types of finish which are applicable to stainless steel. The principal types are mill finishes, polished finishes, electrochemical finishes, and textured finishes. The polished finishes are usually specified for architectural work, and among these the dull polished finish, produced by fine grinding with abrasives, is widely used. Polished finishes may be matched accurately, and any accidental damage to them may be repaired. It is normal to protect the finish during building operations.

Because stainless steel has a finish which is more or less reflective, optical effects have been associated with its use. For this reason, large flat areas should be avoided wherever possible, in order to minimize the possibility o optical distortion effects on the building surface. There are a number of alternative ways of avoiding this effect including the use of ribbed and textured finishes.

Under normal conditions, stainless steel does not cause any galvanic corrosion of other metals, nor does stainless steel suffer any attack from this cause. Under very aggressive conditions or in marine situations, stainless steel might tend to promote corrosion in the metals aluminium and zinc to which it is particularly cathodic. Under such circumstances trouble is avoided by the use of insulating washers or bituminous or zinc chromate paint. The general absence of bimetallic corrosion problems makes stainless steel fastenings particularly useful.

All of the standard methods of jointing may be used with stainless steel. The welding processes which are used with the 18/8 and 18/10/3 steels would normally be carried out by the manufacturers. Stainless steels can loose their passivity and hence corrode around welds because of the chromium having formed chromium carbides during the heating processes of welding. This problem is overcome by the use of either special low-carbon stainless steels, or stainless steels containing stabilizing elements. Welding of stainless steels should, therefore, only be carried out after consultation with the manufacturers.

The many applications of stainless steel externally on shop fronts, window framings, and curtain wall systems, over almost forty years, indicate that these alloys have an outstanding performance. Provided that the appropriate grade of stainless steel is chosen and the precipitated grime is removed by washing, the metal should remain bright and lustrous indefinitely.

NOTE ON METRIC EQUIVALENTS. *Throughout this chapter both imperial and metric units have been used. Where published figures have been quoted these have been given in the units used in the original source, and the metric or imperial equivalent quoted in parentheses after these. Such metric or imperial equivalents have been appropriately rounded to an equivalent number of decimal places.*

During this period of changeover a range of units of measurement for the thickness of metal sheet and strip are in use. The imperial units include the inch, the Standard Wire Gauge (SWG), the Zinc Gauge (ZG), and the system used for lead in which the thickness is defined in terms of the weight per square foot of surface. The metric equivalents are essentially defined in BS 4391, and for most of the sheet-metal products referred to in this chapter the thicknesses are likely to be rounded to the nearest multiple of 0·1 mm on metrication. The lead weight gauge is replaced (BS 1178: 1969) by a gauge based upon a series of BS code numbers which correspond to the weights in pounds per square foot.

REFERENCES

1. INSTITUTION OF STRUCTURAL ENGINEERS (1962). *Report on the structural uses of aluminium* (London).
2. AMERICAN SOCIETY FOR TESTING MATERIALS (1956). Symposium on atmospheric corrosion of non-ferrous metals', *ASTM STP* 175.
3. AMERICAN SOCIETY FOR TESTING AND MATERIALS (1968). Metal corrosion in the atmosphere', *ASTM STP* 435.
4. CARTER, V. E. (1968), 'Atmospheric corrosion of aluminium and its alloys: results of six-year exposure tests', *ASTM STP* 435, pp. 257–70.
5. MCGEARY, F. L., SUMMERSON, T. J., AND AILOR, W. H., JR. (1968). 'Atmospheric exposure of non-ferrous metals and alloys—aluminum: seven-year data'. *ASTM STP* 435, pp. 141–74.
6. MATTSSON, E., AND LINGREN, S. (1968). 'Hard-rolled aluminum alloys', *ASTM STP* 435, pp. 240–56.
7. ALUMINIUM FEDERATION (1966). 'Church cupolas', *Aluminium in service— a series of case-histories*, no. 50 (London).
8. EVERETT, L. H. (1962). 'The compatibility of some aluminium alloys with wet building mortars and plasters—where precautions are needed against corrosion', *The Builder*, **202** (6202), 669–72.
9. JONES, F. E., AND TARLETON, R. D. (1963). 'Effect of embedding aluminium and aluminium alloys in building materials', *National Building Studies Research Paper* 36 (London: HMSO).
10. ANDERSON, E. A. (1956). 'The atmospheric corrosion of rolled zinc', *ASTM STP* 175, pp. 126–34.
11. DUNBAR, S. R. (1968). 'Effect of one per cent copper addition on the atmospheric corrosion of rolled zinc,' *ASTM STP* 435, pp. 308–25.
12. TRACY, A. W. (1956). 'Effect of natural atmospheres on copper alloys: 20 year test', *ASTM STP* 175, pp. 67–76.
13. THOMPSON, D. H., TRACY, A. W., and FREEMAN, J. R., JR., (1956). 'The atmospheric corrosion of copper. Results of 20 year tests', *ASTM STP* 175, pp. 77–87.
14. THOMPSON, D. H. (1968). 'Atmospheric corrosion of copper alloys', *ASTM STP* 435, pp. 129–40.
15. MATTSSON, E., and HOLM, R. (1968). 'Copper and copper alloys', *ASTM STP* 435, pp. 187–210.
16. HIERS, G. O., and MINARCIK, E. J. (1956). 'The use of lead and tin outdoors', *ASTM STP* 175, pp. 135–40.

17. THOMAS, H. E., and ALDERSON, H. N. (1968). 'Corrosion rates of mild steel in coastal, industrial and inland areas of northern California', *ASTM STP* 435, pp. 83–94.
18. BRIGGS, C. W. (1968). 'Atmospheric corrosion of carbon and low alloy cast steels', *ASTM STP* 435, pp. 271–84.
19. GILBERT, P. T. (1953). 'The effect of impurities in the metal on the rate of corrosion of zinc and galvanized coatings in the atmosphere', *J. Appl. Chem.*, **3**, 174–81.

BIBLIOGRAPHY

British Standards Institution	BS 1615: 1961	*Anodic oxidation coatings on aluminium.*
	BS 3987　1966	*Anodized wrought aluminium for external architectural applications.*
	BS 2855: 1957	*Corrugated aluminium sheets for general purposes.*
	BS 3428: 1961	*Troughed aluminium building sheet.*
	BS 1470: 1963	*Wrought aluminium and aluminium alloys for general engineering purposes. Sheet and strip.*
	BS 1126: 1957	*General recommendations for the gas welding of wrought aluminium and aluminium alloys.*
	BS 3451: 1962	*Testing fusion welds in aluminium and aluminium alloys.*
	BS 2997: 1958	*Aluminium rainwater goods.*
	BS 849: 1939	*Plain zinc sheet roofing.*
	BS 729: ——	*Zinc coatings on iron and steel articles.* 729: Part 1: 1961　*Hot-dip galvanized coatings.* 729: Part 2: 1961　*Sheradized coatings.*
	BS 1706: 1960	*Electroplated coatings of cadmium and zinc on iron and steel.*
	BS 3382: ——	*Electroplated coatings on threaded components.* 3382: Part 1: 1961　*Cadmium on steel components.* 3382: Part 2: 1961　*Zinc on steel components.*
	BS 2989: 1967	*Hot-dip galvanized plain steel sheet and coil.*
	BS 1431: 1960	*Wrought copper and wrought zinc rainwater goods.*
	BS 2870: 1968	*Rolled copper and copper alloys, sheet, strip and foil.*
	BS 1569: 1965	*Copper sheet and strip for roofing and other building purposes.*
	BS 1878: 1952	*Corrugated copper jointing strip for expansion joints (for use in general building construction).*
	BS 1178: 1969	*Milled lead sheet and strip for building purposes.*
	BS 2569: ——	*Sprayed metal coatings.* 2569: Part 1: 1964　*Protection of iron and steel by aluminium and zinc against atmospheric corrosion.*

Metals

BS 3189: 1959	Phosphate treatment of iron and steel for protection against corrosion.
BS 3740: 1964	Steel plate clad with corrosion resisting steel.
BS 4391: 1969	Recommendation for metric basic sizes for metal wire, sheet and strip.
BS 1006: 1961	Determination of fastness to daylight of coloured textiles.
BS 3019: ——	General recommendations for manual inert-gas tungsten-arc welding.
	3019: Part 1: 1958 Wrought aluminium, aluminium alloys and magnesium alloys.
	3019: Part 2: 1960 Austenitic stainless and heat-resisting steels.
BS 219: 1959	Soft solders.
BS 1494: 1951	Fixing accessories for building purposes
	1494: Part 1: 1964 Fixings for sheet, roof and wall coverings.
	1494: Part 2: 1967 Sundry fixings.
BS 743: 1966	Materials for damp proof courses.
BS 1202: ——	Nails.
	1202: Part 1: 1966 Steel nails.
	1202: Part 2: 1966 Copper nails.
	1202: Part 3: 1962 Aluminium nails.
BS 1210: 1963	Wood screws.
BS 747: 1968	Roofing felts.
BS 4016: 1966	Building papers (breather type).
BS 1521: 1965	Waterproof building papers.
BS 4147: 1967	Hot applied bitumen based coatings for ferrous products.
BS 3634: 1963	Black bitumen oil varnish.
BS 3416: 1961	Black bitumen coating solutions for cold application.
BS 2717: 1956	Glossary of terms applicable to roof coverings.
CP 143: ——	Sheet roof and wall coverings.
	143: Part 1: 1958 Aluminium, corrugated and troughed.
	143: Part 2: 1961 Galvanized corrugated steel.
	143: Part 3: 1960 Lead.
	143: Part 4: 1960 Copper.
	143: Part 5: 1964 Zinc.
	143: Part 7: 1965 Aluminium.
CP 2008: 1966	Protection of iron and steel structures from corrosion.
CP 231: 1966	Painting of buildings.
PD 420: 1953	Methods of protection against corrosion for light gauge steel used in building.

Building Research Station BRS Digest, nos. 110 and 111 (Frst Series), 'Corrosion of non-ferrous metals'

BRS Digest nos. 29 and 30 (Second Series), 'Aluminium in building: 1: Properties and uses; 2: Finishes'.

BRS Digests, nos. 70 and 71, 'Painting metals in buildings: 1: Iron and steel; 2: Non-ferrous metals'.

BRS Current Paper, Design Series, no. 62, Everett, L. H., and Tarleton, R. D., 'Recognition of corrosion hazards to metals in buildings'.

In addition to the advice contained in the British Standards and in the Codes of Practice and in other sources, a great deal of detailed information is available on the use of both ferrous and non-ferrous metals in buildings from the appropriate development associations, and through the publications produced by these organizations. These associations include:

Aluminium Federation—Portland House, Stag Place, London, SW 1.

Copper Development Association—55 South Audley Street, London, W 1.

Lead Development Association—34 Berkeley Square, London, W 1.

Zinc Development Association—34 Berkeley Square, London, W 1.

Stainless Steel Development Association—7 Old Park Lane, London, W 1.

British Non-Ferrous Metals Research Association—Euston Street, London, NW 1.

British Iron and Steel Research Association—24 Buckingham Gate, London, SW 1.

Chapter 6 PLASTICS

KENNETH A. SCOTT
BSc, ARIC, FPI
*Manager, Plastics Technology Group,
Rubber and Plastics Research Association
of Great Britain*

INTRODUCTION

In the first chapter of this book a summary was given of the factors which influence the weathering and durability of building materials. The fundamental relationship between the design, workmanship, environment, and use of the materials was described. Criteria for judging the performance of a building material were also considered. All of these factors quite clearly will apply to building components which are made from plastics materials. However, the range of available plastics is so wide that a careful selection has to be made to ensure that the material used is the most suitable for a particular application.

For plastics to be used successfully in building they have to offer advantages over traditional materials. Essentially they must fulfil the performance requirements of the component economically. Plastics used in building are usually strong and durable and lend themselves to industrialized manufacturing techniques. They can be easy to instal because of their light weight and resilience.

Maintenance costs of the component can also be reduced. Plastics are often available in a range of colours and allow for styling and colour effects in the exterior decoration of buildings. An example of a successful use of plastics is in the PVC rainwater gutters and down pipes. These are manufactured by a continuous extrusion process and the fittings can be quickly moulded. The resilience of plastics permit the gutter to be sprung into the supports and since they are lightweight, installation is relatively simple. Admittedly the range of colours available is not great but the self-coloured product requires only little maintenance.

PVC has already been mentioned, and it is desirable at this stage to emphasize that it is inappropriate to generalize on the properties of 'plastics'; instead the individual polymers have to be considered, since not all plastics are suitable for exterior use.

'Plastics' is the generic term for an arbitrary group of materials based on synthetic or modified natural polymers which at some stage of manufacture can be formed to a shape by flow, aided in many cases by heat and pressure. This wide range of materials can be classified into two major groups, namely thermoplastics and thermosets. Representatives of both types are used in building. Thermoplastics are capable of being softened by heat and hardened by cooling. These melting and freezing phenomena are repeatable. Thermosetting materials on the other hand, after moulding do not soften significantly on heating to temperatures below the decomposition temperature.

A wide range of plastics is available with a considerable spectrum of physical properties. Many of these unique properties have been applied to almost every industry and technology ranging over land, sea, air, and space transportation, electronics, dentistry, surgery, domestic appliances, furniture, chemical plant, etc. The building industry consumes about 20 per cent of the plastics manufactured in the United Kingdom i.e. about 220 000 tons in

Plastics

1969. For exterior applications the most important materials to consider are:

Polyvinyl chloride (PVC):
    Cladding   5000 tons in 1969.
    Rainwater goods   12 800 ,,   ,,
    Roof lights   5500 ,,   ,,
    Surface coatings
    Window frames

Glass-fibre reinforced polyester resins:
    Cladding   1300 tons in 1969.
    Roof lights   5300 ,,   ,,
    Window frames
    Decorative effects

Phenol/formaldehyde—melamine formaldehyde laminates:
    Decorative cladding

Acrylic sheet:
    Glazing—roof lights
    Cladding
    Decorative effects

Acrylonitrile butadiene styrene (ABS)

Polyvinyl fluoride

Acrylic surfacing film

It must be borne in mind that each of these polymers can often be formulated in order to change some of their specific properties. For example, stabilizers can be added to improve the light stability of a compound, additives can be used to improve the fire performance and many of the physical properties, such as water-absorption and abrasion-resistance. Light transmission and rigidity, opacity, and so on can also be adjusted. Some of the most spectacular adjustments of properties can be achieved with PVC. Thus one can have rigid, opaque PVC for rainwater goods, pressure pipes, cladding, etc., or transparent PVC sheet for roofing purposes and skylights. Flexible formulations or plasticized formulations of PVC are also used for the manufacture of leather cloth and electrical insulation of wiring and cables. PVC can also be produced in an expanded form for thermal insulation, whilst other formulations are used as surface coatings. Some caution is therefore needed to ensure that the grades or types within the generic plastic material are specified. Thus some grades of PVC will be totally unsuitable for use in exterior applications whilst others may have a life of thirty years or more.

FACTORS WHICH INFLUENCE THE WEATHERING AND DURABILITY OF PLASTICS MATERIALS

Weathering effects are caused by a number of interacting factors which comprise weather. The dominant ones are ultra-violet radiation, temperature, water, oxygen and micro-organisms, industrial gases, and mechanical loadings which arise through exposure to wind and snow loadings (1, 2). All materials are likely to be affected by these components but may well be influenced to different degrees by the individual components. Plastics have a very low water-absorption and therefore are not damaged to a great degree by exposure to water, unless, of course, the polymer is one which can be hydrolysed and even under these conditions a combination of water with heat or radiation is necessary to cause damage. Natural radiation consisting of ultra-violet, visible, and infra-red rays has a major effect on a number of plastics materials, particularly if they are not specially stablized. The effects can arise from a chemical change taking place within the polymer sometimes giving rise to a yellow discoloration and often to a development of brittleness. Care is taken in the formulation of the resin to minimize these effects. Most plastics which are used for exterior applications contain a suitable ultra-violet absorber (3, 4) The influence of these additives can be spectacular, for example, polyethylene which is not normally used widely in exterior applications, can become extremely brittle after exposure to ultra-violet light for a year or two. If, however, the polyethylene is pigmented, filled, or contains carbon black (ultra-violet absorber), then its stability is greatly increased and a service life of many years is possible (5, 6, 7).

Thermoplastics will soften if they are hot enough and it is important, therefore, to ensure that a thermoplastic component is not exposed to too high a temperature, otherwise distortion will occur. Most data sheets on thermoplastics indicate the heat distortion temperature of the material and this can serve as a guide to its suitability. It is important to note that the colour of the moulding can influence the temperature to which the surface may rise if the component is exposed to direct sunlight. These high temperatures can accelerate the oxidation and thermal degradation of some polymers unless suitable stabilizers are added (7).

Some polymers can become brittle at low temperatures. This is a simple thermal effect and the original strength and toughness is restored at normal temperatures.

Oxidative degradation accelerated by high ambient temperatures, ultra-violet light, and moisture can result in a change in the molecular structure of the polymer and a consequent loss of strength (8).

These effects of ultra-violet light, heat, and rain can occur to varying extent with most polymers. Their effects are negligible on polyvinyl fluoride, acrylics, phenol formaldehyde, melamine formaldehyde, and suitably formulated polyvinyl chloride and polyester resins.

In most cases plastics materials are reasonably resistant to industrial atmospheres and attack by micro-organisms (9). This is not to say they do not become discoloured, dirtied, or covered with an unsightly deposit. More often than not a simple cleaning action will remove this deposit and expose the undamaged polymer surface.

POLYVINYL CHLORIDE (PVC)

Polyvinyl chloride is the polymer which is used most extensively in the building industry. It is well known in applications such as wall and floor coverings, electrical insulation and conduit, roofing panels, rainwater guttering and down pipes, soil and waste pipes, cladding, and recently its use in window frames has been developed.

It is an extremely versatile polymer in that it can be formulated to give rigid or flexible products which may be transparent or opaque. It is available in expanded forms for thermal insulation. Pastes can be formulated for metal dipping and surface coating. Products can be manufactured by extrusion, calendering, injection moulding, vacuum forming, welding, rotational and blow moulding techniques. For most of the exterior building applications extruded PVC is used, fittings are injection moulded whilst special techniques are used for coating mild steel plate. In most cases rigid PVC formulations are used.

Unstabilized PVC is notoriously unstable (10, 11) and on exposure to heat and light it rapidly discolours and a serious embrittlement is apparent. Indeed the technology of PVC would not have progressed at all had it not been for the development of effective stabilizers and processing aids. It would not be appropriate to consider the details of PVC formulation in this paper, but an idea of the principles involved can be helpful.

The PVC polymer as manufactured is a powder which as has already been said will degrade at the temperatures needed for fabrication. Stabilizers are added to overcome this, the most usual ones are lead compounds such as tribasic lead sulphate or dibasic lead stearate. Transparent panels employ stabilizers based on tin. During processing where high shear forces are applied, excessive heat can be generated—this can be controlled and processing aided by the addition of lubricants. These are waxy materials used in relatively low concentrations. The surface gloss of an extruded product is often determined by the lubricant or mixture of lubricants used in the formulation.

For exterior applications ultra-violet absorbers are used to minimize discoloration by solar radiation.

In addition to these essential components of the formulation, fillers are sometimes used to improve impact strength. In other formulations impact modifiers such as acrylics are used.

PVC is self-extinguishing but its performance in this respect can be improved by the addition of antimony trioxide. In some building applications such additions are desirable.

The weathering properites of PVC will depend on the formulation used and on the conditions under which the components are moulded. Excessive heating during moulding can seriously undermine the durability of the product. The principles of formulation and processing which determine durability are generally known and are well established. Users of PVC components should obtain firm assurances from their suppliers regarding the weathering characteristics of the product. It can be reasonably argued that in this highly competitive field reputable manufacturers will ensure that the quality of their product is as high as present-day technology will allow.

It is sometimes difficult, from published data, to determine the life which can be anticipated for PVC products. This is because of the range of PVC formulations which are possible. Although PVC is considered to be one of the most weather-resistant polymers available today, the opinions of a number of leading experts in the field make very interesting reading. Penn (12) states that PVC is weather-resistant whilst Plumb (13) says that a 15–20 year guarantee on suitably formulated unplasticized PVC products can be issued on a reasonable risk basis. Crowder (14) states that rigid PVC rainwater goods would probably be mechanically sound for thirty years but that transparent PVC products would lose their light transmission properties after a period of only ten years. DeCoste and Walder (15) however, hold the view that the weatherability must be improved. Thacker and Nass (16) feel that something has to be done to retain physical and mechanical properties of PVC. Leeper and Gomez (17) say that the formulation compounding and processing have to be critically controlled in order to obtain good weatherability in a PVC product. DeCoste and Hansen (18) say that colorants could improve the life of a PVC product; these comments lead to a statement by DeCoste, Howard, and Walder that PVC is not inherently weather-resistant and it derives this property mainly from its formulation. Estevez (19) says the important thing is not how resistant the polymer is to weather but how resistant it can be made by formulation. He is of the opinion that weatherability of a PVC product is the joint responsibility of the technologists of the raw material suppliers and the fabricators. However, some raw material suppliers now tend to disagree with this opinion as they feel that as many of the fabricators carried out their own formulation and compounding work the responsibility for the durability of the product rests with them.

These comments only go to emphasize the important role which the processor has in producing PVC components with good weathering characteristics. Apart from the selection and correct compounding of a suitable formulation, the processor can influence the properties of the final product in two major ways: *a*, during the processing the polymer may become excessively heated, this causes a yellowing and an embrittlement of the product, and *b*, the processing conditions may also be insufficient to obtain a good compounding and fusion of the component and this can again influence the mechanical and durability properties. During extrusion it is also possible to orientate the molecules of the compound so that there is a grain effect in the mechanical properties of the material. This often results in a product having a lower impact strength than is expected. Clearly the fabricator has a responsibility to ensure that the carefully formulated and durable compound is converted into a similarly stable component.

Case-histories Recently the Rubber and Plastic Research Association (RAPRA) has made a pilot case-history study of the durability of plastics in exterior building applications. This was part of a larger study on weathering which is sponsored by the British Plastics Federation's Building Group. A number of lessons can be learnt from these studies—one of the most important was the need for close attention to fixing and to architectural detailing.

PVC-coated wire fencing was introduced round about 1954 and has proved to be extremely satisfactory. Rigid PVC has been used satisfactorily in outdoor directional signs and road signs for over fifteen years. In the late 1950s wall and roof claddings were introduced and about 1960 translucent sheet PVC came on to the market. In 1958 mains water piping for services were introduced and now BS 3505 covers this application. In 1958 PVC gutters and down pipes became available. The change and development in this particular application during the last decade indicates how alive PVC fabricators and raw material suppliers and designers have been to improve the quality and design of the gutters and down pipes to suit the requirements set down by the building industry. It is therefore not surprising that PVC sections now hold well over 70 per cent of the market in the UK. About 1958 window frames made from PVC were introduced on the continent. PVC soil systems have been available since 1960 and have been proved very efficient. Other PVC products which have been on the market for some time and have shown their merit are floor coverings, electrical conduit and internal wall claddings.

Coloured plasticized PVC film on metal sheeting as a protective and decorative finish, has been used to good effect in many outdoor applications. An automatic car park at Southwark Bridge clad in 1963 (Fig. 6.1) with PVC coated steel sheeting, was recently found to have retained much of the original colour although this is partly obscured by dirt from the industrial atmosphere. Such a coating may harden with age, through loss of plasticizer, and the stability of the brighter colours may leave something to be desired, but protection of the metal is likely to remain adequate for many years, particularly if proper attention is paid to cut edges and holes which are the main points of weakness. A certain patchiness in colour which has arisen from an uneven cleaning of the panels can be seen. It is interesting to note a rust stain (Fig. 6.2) at the base of these columns which could have been avoided by means of improved detailing and allowance for drainage channels at the base of these columns.

Rigid PVC rainwater goods provide the most widespread evidence of the weathering behaviour of this material. After some early experience of a range of colours, availability is now limited to black, white, or grey products which do not fade as did many of the early ones. Some greys may still change in appearance after long periods of exposure but this may well be tolerable. Painting to improve appearance does not call for any special precautions except cleaning. In many cases fading of colour is acceptable, provided that the change of colour is consistent. Processing and the selection of an adequately formulated compound can influence the amount of colour change.

Fig.6.1. PVC coated steel columns.

Fig.6.2. Rust staining at the base of PVC-coated steel columns.

Fig.6.3. Weathered PVC down pipe, gutters, and fittings.

Photograph by courtesy of the Building Research Station.

Fig. 6.3 shows the different behaviour of injection-moulded fittings and extruded pipes and gutters in an experimental product which was made in the very early stages of development of these materials. The injection-moulded component has changed very little in appearance whilst the extruded part with different stabilizer and lubricant systems has faded over the course of years. The streaky effect in this fading is due to a peculiarity of the processing condition.

An attraction of the PVC rainwater system is that brackets and fittings can be made of the same material and therefore will be as durable as the rest of the product. The indications are that PVC rainwater goods will give a good service for twenty years or more, although there is some loss of resistance to impact, and discoloration can occur. The principal weakness on weathering is often deterioration of the synthetic rubber gaskets used to provide waterproof joints and accommodate thermal movement.

In 1952 a directional sign made of black and white rigid PVC laminate was fitted outside the Royal Infirmary at Salford. The sign was removed after ten years because

Plastics

the message conveyed needed alteration and it was returned to the PVC manufacturers. Close examination showed it to be in excellent mechanical condition with almost no deterioration. It was greyish due to a coating of airborne dirt, which was removed by simple washing to reveal the original surface of the sheet unmarked by ingrained dirt, crazing, chalking, or pitting and with only the merest trace of yellowing. Although PVC is capable of such good weathering behaviour, careful formulation and manufacture is necessary to ensure that it is achieved.

Corrugated translucent or transparent PVC sheet used for roof lighting does not weather as well as the opaque form of the material because of the greater difficulty in stabilization and exposure to ultra-violet light. The opacity of the sheet increases on exposure and eventually reduction in light transmission may necessitate replacement. The consequent colour changes frequently range

Fig.6.4. Weathered PVC transparent sheet. The dark areas are those which were in contact with supports.
Photograph by courtesy of the Building Research Station.

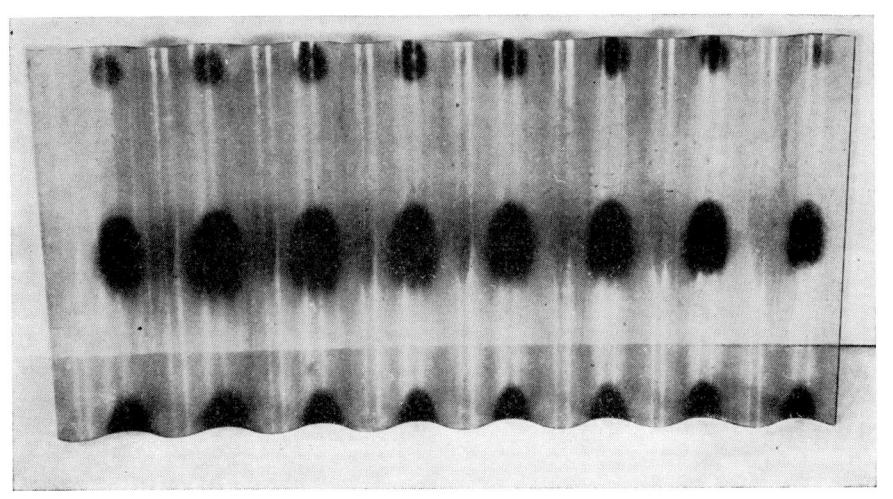

from yellow to a dark brown. In such panels a combination of heat and ultra-violet light cause the greatest deterioration.

A dramatic effect is illustrated in Fig. 6.4. The PVC test sheet was exposed on a weathering rack in such a way that parts of the sheet were in contact with a wooden cross-member. Because air could not circulate freely in this region the polymer was, on average, hotter in this region, consequently a dark discoloration occurred. Care has to be taken in installing components to prevent an image of the supports arising in this way.

GLASS-FIBRE REINFORCED POLYESTER RESIN (GRP)

Glass-fibre reinforced polyester resin laminates are complex structures and an appreciation of their weathering characteristics cannot be gained without a consideration of the chemistry of the polyester resins, the nature of the glass reinforcement, and the influence of the structure of the composite and any additives it contains. GRP in the past has been considered to be a wonder material; it has been used extensively in aircraft, boats, chemical plant, swimming pools, water tanks, many of the advanced prefabricated housing systems, church spires, etc. Often it has had the reputation of being unbreakable and indeed there is a fairly recent publication issued in February 1967 (20) in which playground equipment is considered. The article contains a certain degree of over-enthusiasm but many of the claims are well justified.

Playground equipment must be rugged, attractive and completely safe. No lacerations, splinters, abrasions. It must be able to stand up to not only normal wear and tear but to additional punishment dished out by children. With a few exceptions, it is outside most of the year so it must withstand exposure to sun, wind and rain, possibly even snow and ice.

Moreover such equipment, particularly if colourful, will draw vandals like a magnet. What's called for is a material which is child-proof, weather-proof, vandal-proof and fool-proof. In a nut-shell—reinforced plastics.

However, in considering material for building applications it is important to know the criteria by which quality is to be judged. In building it is important not only that the component maintains its function mechanically but also it must continue to be acceptable visually without excessive maintenance.

In general, cast polyester resin with no reinforcement has excellent weathering properties. Some differences in weathering performance arise from the particular raw materials which are used in the manufacture of the basic resin. Certain resins are flexible and have a higher water-absorption than the general-purpose resins, others by virtue of the modifying acids used in their manufacture may be self-extinguishing and difficult to ignite. Each of these types in general have poorer weathering characteristics than many of the general purpose formulations. Glass, of course, is considered to have excellent weathering characteristics but glass fibre may not be so durable as bulk glass because of its very large surface area. The weathering resistance of the combination of resin and glass fibre is very sensitive to the quality and form of the combination. In particular, if the glass fibre is near the surface of the laminate a rapid deterioration of appearance can occur on weathering. This is because the resin may erode from the surface fibres and cause them to become prominent and exposed. This often gives a centre for the collecting of dirt and frequently a whisker-like effect is obtained.

It is not appropriate in this book to consider in detail the chemistry of the polyester resins, but a brief outline of the principles may be helpful (21, 22). The polyester resin consists of a number of basic ingredients or components. In the first instance we have a polyester which is produced by the reaction between a glycol and a dibasic acid or from mixtures of various types of glycol and dibasic acid. The specific characteristics of the resin are determined by the components which are used. Thus, for example, a resin containing diethylene glycol is likely to have a greater toughness but a higher water-absorption than one containing predominantly propylene glycol. Phthalic acid is the modifying acid which is most frequently used in the general purpose resins. The weathering characteristics of such materials are reasonably good. Very often the phthalic acid component can be replaced by a material hexachloroendo methylene tetrahydrophthalic acid or more simply 'HET' acid which contains a substantial quantity of chlorine atoms. This acid has the very desirable characteristic of reducing the flammability of the resin to produce materials which are capable of achieving EXT. SAA classifications or even Class 1 Spread of Flame grading according to BS 476.[1] Unfortunately, the use of HET acid reduces the weathering properties of the resin (23).

The reaction product between the glycol and the dibasic acid is a polyester which can be extremely viscous syrup or possibly a solid. This is then dissolved in styrene monomer which is a material of low viscosity and solvent-like in its properties in which the resulting solution is a thin syrup. In many resins designed for translucent panels, methylmethacrylate monomer is also used (24, 25). This thin syrup needs to be stabilized to prevent premature gelation, frequently hydroquinone is used for this purpose. The attraction of this type of polyester resin is that it can be activated by the addition of a peroxide catalyst and an accelerator and will cross-link to form a thermoset product which is fusible. The resin described above would

---

[1]. The EXT SAA classification is the highest given in BS 476. Part 3: 1958. It indicates that the roofing panel has not been penetrated by flame within one hour and no spread of flame has occurred on the product under the conditions of test.

be the basic type used in general laminating work (26). Such a resin if exposed to ultra-violet light will become more yellow in colour; this would be undesirable for normal building applications. However, additions of a ultra-violet absorber will prevent this discoloration occurring to any marked extent even after a long period of exposure to natural weathering (4, 28, 29). In the manufacture of opaque laminates it is frequently the practice to add a filler, such as calcium carbonate, a clay, or a powdered silica to the resin. Some additives are very helpful in improving and facilitating production methods and enhancing some mechanical properties of the completed laminate. Most general-purpose laminates or resins used in the building industry are likely to contain additives. The resins can also be pigmented. The colour stability of the pigmented laminate will often depend more on the pigment which is used than on the resin, although quite clearly a resin without a ultra-violet stabilizer is likely to impart a yellow discoloration in the composite.

The polyester resin described above, when it is fully cured or hardened is not strong enough for structural uses, and so it has to be reinforced with glass fibre. The degree of reinforcement achieved depends on the quantity and orientation of the glass fibre used. This is illustrated in Table 6.1.

Table 6.1. *Flexural strength of glass reinforced polyester resin*

Type of reinforcement	Quantity of glass fibre (percentage by weight of laminate)	Flexural strength N/mm^2
Unidirectional-filament wound	50–75	680–980
Woven fabric—satin weave	50–70	350–520
Woven roving	45–60	200–300
Chopped strand mat	25–40	130–300

Fig.6.5. Weathered GRP, showing the glass fibre pattern.
Photograph by courtesy of the Building Research Station.

In most building applications chopped strand or swirled mat reinforcements are employed. The quality of the glass-fibre reinforced laminate depends to a large extent on the way in which it is constructed. An appreciation of this can be obtained by studying the mechanism by which serious deterioration can take place (24, 25, 30).

Fig. 6.5 shows a weathered panel in which silvery glass fibres can be seen at the surface. This effect is obtained only if fibres are actually on the surface of the laminate. Although this may not have been obvious when the moulding was first made. Consider the mechanism of the exposure of fibres as shown in the drawings.

In the original moulding (Fig. 6.6) the fibres which are near to the surface are coated with a thin layer of resin. On repeated exposure to rain and sunlight (heat) the bundles of fibres may swell and cause the thin resin surface to craze. On drying and loss of moisture the forces at the surface will be reduced. Repetition of this cycle over a period will lead to an erosion of the thin

Plastics

Fig.6.6.

A sketch of an unweathered grip surface in which the fibres are near the surface.

Fig.6.7.

A laminate of the type in Fig. 6.6.

Fig.6.8.

Glass fibres protected by a gel coat.

Fig.6.9.

Glass fibres 'proud' on the surface due to excessive compression by flexible film.

layer of resin covering the surface fibres, which then break through at the surface. Subsequent weathering Fig. 6.7 will aggravate the situation and the fibres will become more and more exposed until individual filaments in the fibre bundles may break away and produce a whiskery effect.

The resulting surface is then one on which dirt can become embedded and the result is unattractive. This mechanism can be readily visualized by an examination of the stereoscan micrographs produced by the Building Research Station (Fig. 6.10–6.12).

This mechanism of deterioration of fibre-reinforced laminates depends on a failure of the very thin layer of resin covering fibres at the surface of the laminate. It can clearly be avoided by protecting the fibres by a surface coating or by ensuring that they are below the surface of the laminate. This in the opinion of the author is the most important requirement for durability in GRP.

The most widely used moulding technique used for

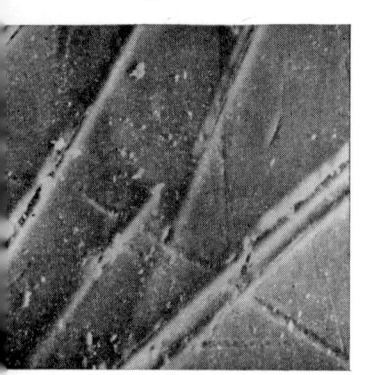

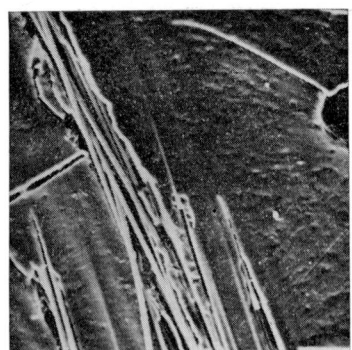

Figs.6.10–6.12. Stereoscan electron-micrographs showing the progressive exposure of glass fibres on weathering.

Photographs by courtesy of the Building Research Station.

GRP is the so-called hand lay-up process (26). This is a relatively simple operation in which adequate protection of surface fibres can be readily achieved even in the most complex shapes. A 'female' mould is used—that is, the mould is such that the exterior or functional surface of the moulding is made in contact with the mould surface. The procedure is to treat the mould surface with a release agent and then to coat it with a gel coat resin. This is a polyester resin which is formulated to be tough and durable. The resin is catalysed so that it will cure or harden at ambient temperatures. It is 'painted' evenly over the mould surface and when it is hardened a laminate of resin and glass fibre is applied and 'consolidated' by well-established techniques. When the whole moulding has hardened it may be heated to obtain full cure before being removed from the mould. The gel coat is now an integral part of the moulding and provides a resin rich surface which protects the bundles of glass-fibre strands in the reinforcement (Fig. 6.8).

Mouldings which are produced in large quantities may be made by hot-pressing techniques. In these processes it may not be possible to use a gel coat, and the process is such that glass fibres can be near to the surface of the moulding. In such products it is usual to coat the surface of the moulding with a polyurethane resin suitably formulated for durability.

Translucent roofing panels are usually manufactured by a continuous process in which curing or hardening of the resin is accelerated by heating. Resin-impregnated glass fibre is sandwiched between regenerated cellulose film and the sandwich is fed through rollers to remove air bubbles and to 'consolidate' the laminate. In doing this there is a risk of increasing the concentration of fibres at the surface as illustrated in Fig. 6.9. This is because the flexible cellulosic film follows the contours of the glass-fibre reinforcement to a limited extent and sometimes the surface fibres can be very slightly proud of the surface—it is as if they have been pressed into the surface from inside the laminate. The problem can be minimized by using a surfacing tissue or putting a thin gel coat or protective layer on the cellulosic film before the main laminate is applied. Such techniques have been described by Brindley (31).

Possibly the most convenient method of applying a protective surface, however, is to spray with a polyurethane or acrylic-based lacquer.

The problem of surface fibres can be minimized by using specially designed mat reinforcement in which fine fibre strands are used and in which the mat is tightly bound as opposed to loosely bound. To obtain panels with a high transparency resins containing methyl methacrylate monomer are used. These resins not only give better clarity than general purpose resins but also seem to wet and impregnate the mats more efficiently. As a result translucent panels made with resins containing methyl methacrylate monomer have better weathering characteristics than similar products without this monomer.

For many applications in building it is necessary to use panels which have some degree of flame-resistance, i.e. they need to obtain a reasonably good classification within a BS 476 test. Glass-fibre reinforced polyester laminates,

Table 6.2.1 *Effect of gel coats and surfacing tissue on gloss retention (32).*

	Percentage gloss retention after exposure for (years)			
Type of surface	0	2	4	5
Gel coat	100	98	97	97
Terylene tissue	100	95	92	92
Glass-fibre tissue	100	93	83	78
No gel coat	100	75	30	12

Table 6.2.2 *Effect on light transmission (33).*

	Percentage light transmission		
Type of surface	Before exposure	After exposure*	Percentage loss
No gel coat	82	70	12
With surface tissue	78	76	2

* Period of exposure not recorded.

suitably formulated can obtain EXT. SAA classification by BS 476:Part 3: *Fire performance of roofing Panels* or Class 1 Spread of Flame by BS 476: Part 1: 1953. Unfortunately, these resins having improved fire performance often have weathering characteristics which are not as good as a suitably formulated non-self-extinguishing resin. However, surface protection with suitable gel coats or polyurethane coatings overcome this difficulty.

The surface protection methods described above overcome deterioration of laminates in which fibres 'pop out' of the surface. Clearly these surface coatings must themselves be durable and retain their colour. The literature indicates methods of improving the performance of polyester resins for gel coats. Recommendations include the use of ultra-violet absorbers, and apart from pigments it is suggested that the use of fillers should be minimized in the gel coat since they can increase the loss of gloss and fading of colour on ageing (32, 34).

Plastics

Table 6.3.1 *Effects of monomer on weathering performance (25). After three-year exposure.*

Composition	Gloss retention	Fibre 'pop out'
Polyester/methyl methacrylate/styrene		
75/25/0	54·5	Very bad
75/0/25	83·0	Bad
60/20/20	82·5	Slight
60/10/30	62·5	Some
60/0/40	78·1	Bad

Table 6.3.2 *Effects of monomer on light transmission after weathering exposure (32).*

	Percentage light transmission after (years)			
	0	2	4	5
Styrene monomer only	75	67	60	55
1 : 1 Styrene methyl methacrylate	82	80	74	69

The surface of the gel coat can influence the durability of panels. A highly glossed, flat surface may be easy to maintain by cleaning but it may also show up imperfections. A textured surface could improve durability and conceal imperfections in the surface—in the long term it might also provide recesses which would retain dirt. Rugger has noted that a roofing panel having a 'crinkle' surface finish retains its light transmission characteristics far better than a panel with a plain surface (35).

Some commercial panels are surfaced with granite chippings, pebbles, or ceramics. The main surface presented to weathering is then not resin but the traditional building material, the weathering characteristics of which are well known. The main criterion of durability in these products is the retention of the adhesive bond which holds the chippings on the surface. In general adhesion is good and not adversely affected by weathering.

A recent development in the surface protection of GRP laminates is in the use of polymer films based on polyvinyl fluoride (PVF) or acrylates (35, 36). The film is specially treated to bond to a GRP laminated mould and cured on it. In an accelerated exposure test in the aggressive conditions of Florida it was found that GRP surfaces with PVF maintain their initial gloss and appearance for at least six years whilst conventional roofing panels exposed under the same condition began to show erosion and fibre exposure after only twelve months exposure. Indications are that panels coated with PVF film would require no maintenance for fifteen years.

The degree of cure of the resin used is an important factor in determining the properties of GRP laminates. The higher the degree of cure the more satisfactory will be the laminate. There are no convenient methods of measuring the degree of cure of a laminate with any certainty. It is usual to ensure an adequate cure by accelerating the curing process by heating the moulding. The conditions for heating will depend on the resins used and the curing systems employed. However, when post curing is used care should be taken not to use too high a temperature lest it causes a yellow discoloration (23, 37, 38).

Maintenance (39)

The correct selection of resin and glass-fibre reinforcement and their correct combination in laminating and curing will ensure that the maximum durability is achieved. In such laminates the most likely need for maintenance arises from the accumulation of dirt and it is not unusual to recommend that exposed mouldings should be washed with water or soapy water at regular intervals.

Should the laminates have weathered to a degree where fibre exposure and dirt accumulation is unaccept-

Plastics

able then a procedure for refurbishing is often recommended. The surface should first be thoroughly cleaned, using steel wool to remove the dirt and surface fibres. After drying, the surface is coated with an acrylic varnish (containing an ultra-violet absorber) applied with a brush. This procedure not only recovers a very high proportion of light transmission and surface gloss but also often removes most of the yellow discoloration caused by ultra-violet light. This is because most of the yellowing is at the surface of the product.

Some manufacturers consider that a simple hosing of the exposed product is more than sufficient maintenance for the first seven years. Should refurbishing then be required it is claimed that after treatment the surface should be adequate with similar hosing for a further seven years. When PVF film is used on the surface it is claimed that a simple hosing is all that will be required for the first fifteen years. The manufacturers suggest that since this is a relatively new product the fifteen-year estimated period may well be a pessimistic one.

Case-histories   Having considered the factors which can determine the durability of a laminate one is faced with trying to find an answer to the simple question: How long will GRP panels last? Clearly it is a difficult question to answer since so much depends on the environment, the requirement of the product, and the quality of the original laminate. Some guidance can be obtained from the literature reporting the controlled studies in various parts of the world. For example, Rugger of Piccatinny Arsenal in the USA suggests that the best which one can expect from a GRP roofing panel is about fifteen years of satisfactory performance. Note that this is probably a judgement of the period of adequate light transmission and not of retention of integrity—the panel is not likely to disintegrate after fifteen years. Dr Crowder of the Building

Research Station considers on the basis of panels he has examined that good panels should have a life of thirty years or more in a temperate climate. It is generally believed that well-constructed laminates surfaced with a PVF film will have a longer life than this. The RAPRA case-history study suggests that the estimate of a thirty-year life is reasonable, although it has to be said that no GRP laminates which were thirty years old were seen. This is not generally surprising since GRP, in the United Kingdom, only began its significant commercial development in about 1953.

Many of the deficiencies observed during the survey arose from mechanical failure of the component due to inadequate fixing or to applied forces which were not allowed for in drawing up the performance requirement specification. Other applications emphasized that the criteria by which one panel was judged to be satisfactory were those which proved another to be unacceptable or at least disappointing. For example, roof-light panels fitted to a warehouse at Rochester in 1953 are at present very dirty and discoloured. This is due to a dirt deposit, fibre exposure, and yellowing of the resin, particularly at troughs (Fig. 6.13). There is a considerable dirt and possibly algae accumulation at the overlaps between panels. This roof has received no cleaning or maintenance of any sort during its eighteen-year life. However, the user of the building considers that the roof is functioning sufficiently well for his purpose and that adequate light comes though the panels. Obviously panels in this condition would be totally unacceptable for the canopy of a swimming pool.

Glass-fibre reinforced polyester resin panels have often been sold on the basis that they were unbreakable or vandal-proof. This knowledge seems to be a clear invitation to attack the panels to prove or disprove the point. The challenge appeared to have been accepted at a school which was visited. GRP panels at ground level and facing

Plastics

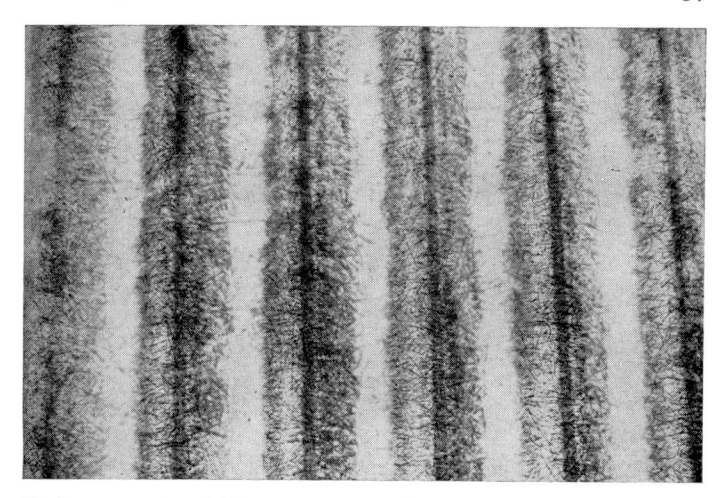

Fig.6.13. Weathered GRP corrugated roofing viewed from inside a building. Note the dirt accumulation and accentuation of fibres in the troughs.

the school playing field had been damaged by the impact of cricket balls, boots, and probably stones.

Methods of fixing can have an important bearing on the performance of the whole component.

Corrugated roofing panels installed at the Wembley sports stadium some six years ago are functioning reasonably well. They have been cleaned at intervals. However, the system as a whole is showing signs of deterioration due to the corrosion of the fixing bolts. The specific reason for this failure is not fully known; it was at first thought that ungalvanized bolts had been used in the fixing operation but possibly other explanations are required. It may be, for example, that water has collected between a fixing washer and laminate and has been unable to evaporate away. This stationary water may have caused the corrosion damage to the bolts.

POLYMETHYL METHACRYLATE

One of the earliest plastics to be used out of doors was polymethyl methacrylate, and during the war it was in considerable demand for cockpit covers of fighter aircraft. Polymethyl methacrylate (PMMA) is a glass-like plastic which is lighter and tougher than glass and is usually supplied in sheet form. The polymer is available in a wide range of colours and has excellent outdoor weathering properties (40). 'Over thirty years' experience in the use of PMMA in buildings is available (41). It has been used as a replacement for glass but the cost is significantly higher than that of glass. Advantages lie in the light transmission properties in the order of 92 per cent for clear panels, good impact resistance, light weight, easy formability, easy handling and installation, and outstanding weathering and durability. In common with many plastics materials PMMA is normally considered to have a low abrasion resistance (27). However, there is some evidence to indicate that while commonly accepted abrasion tests result in a low rating, the performance of PMMA whilst subjected to wind-blown sand and dust is better than the test indicated. This is thought to be due to the ability of the material to yield slightly under impact by the particles.

PMMA products include facia panels, skylights, sun shades, light fittings, interior wall panels, door, bath, and shower enclosures, etc.

Over the years there has been a change in the formulation of cast PMMA and this must be borne in mind when conclusions are drawn from some of the earlier published work and earlier samples. For instance in 1957, a trace ultra-violet absorber was included in the ICI product, this is claimed to have had a threefold effect in improving durability. The newer forms of ultra-violet absorber

Plastics

reduce the tendency of the sheet to yellow slightly on exposure and it reduces the drop in molecular weight which the acrylic sheet undergoes on exposure. A drop in molecular weight modifies the mechanical properties particularly in reducing impact strength (4).

Shaped acrylic components are usually manufactured by heating a cast sheet until it becomes sufficiently soft to form into the required shape. This forming action can sometimes introduce stresses into the polymer. These stresses on ageing can result in a degree of cracking and crazing, although this may take a number of years. The risk of crazing can be minimized by the choice of processing conditions and by annealing the moulding to relieve internal stresses (42-4).

Cleaning and maintenance of the PMMA

Periodic cleaning of PMMA products is recommended by all suppliers of the polymer, since it is claimed that it helps the material to retain its good appearance. Hosing or washing with water or warm water and soap is perfectly satisfactory for regular maintenance. Stiff bristle brushes or hard cloths should never be used since they will tend to scratch the surface. A soft mop or cloth is recommended. The frequency of the cleaning is dependent upon locality, for where atmospheric contamination is high, cleaning should be necessary as often as every six months.

Electrostatic charges on the surface of PMMA can cause dust accumulation and wiping with a cloth to remove the dust could cause the creation of further charges and greater dust build-up. The solution lies in making the surface of PMMA conductive so that static electricity can be discharged. This can be achieved by washing with water but this method is not permanent. Application of a substance which has an affinity for water will have some effect when the relative humidity is greater than 50 per cent. Antistatic solutions containing ionic compounds are effective even under very dry conditions because they are themselves conductors and can remain effective for four years if they are undisturbed. However, these can be removed by washing. They may be of value on vertical surfaces but do not prevent the settling of dust on horizontal surfaces (45).

Acrylics are often used to achieve decorative effects because they are available in a range of transparent and opaque colours. The weathering characteristics of the coloured materials are more dependent on the properties of the dye or pigment than on the colour itself. Colour stability certainly depends on the colour stability of the colorant. The effects of colorants are indicated in a publication by ICI Ltd, (46) which recommends the colours which can be used out of doors for periods of 3–5 years, 5–10 years, and more than 10 years. Colours which are not recommended for exterior use are also listed.

Heat can often accelerate the deterioration of polymers exposed to ultra-violet light. Some pigments or dyes will absorb heat whilst others reflect heat—clearly those colours which absorb heat will promote the thermal degradation of the polymer.

Glass-fibre reinforced acrylic laminates (27)

It was noted in the glass-fibre reinforced polyester section that methyl methacrylate monomer improved the weathering characteristics of polyesters containing styrene. Similar glass-fibre reinforced panels have been produced using acrylic syrups as the binder. The syrups are polymerized after impregnation and moulding of the composite. Tests on these laminates in Florida suggest that they may have a weathering durability which is superior to that obtained from the glass-fibre reinforced polyester materials.

Case-histories

Fig. 6.14 shows acrylic roof lights at Berkeley Square House in London which were installed in December 1938. On inspecting the samples in 1968 it was learnt

Plastics 261

Fig.6.14. Acrylic roof lights after thirty years' service.

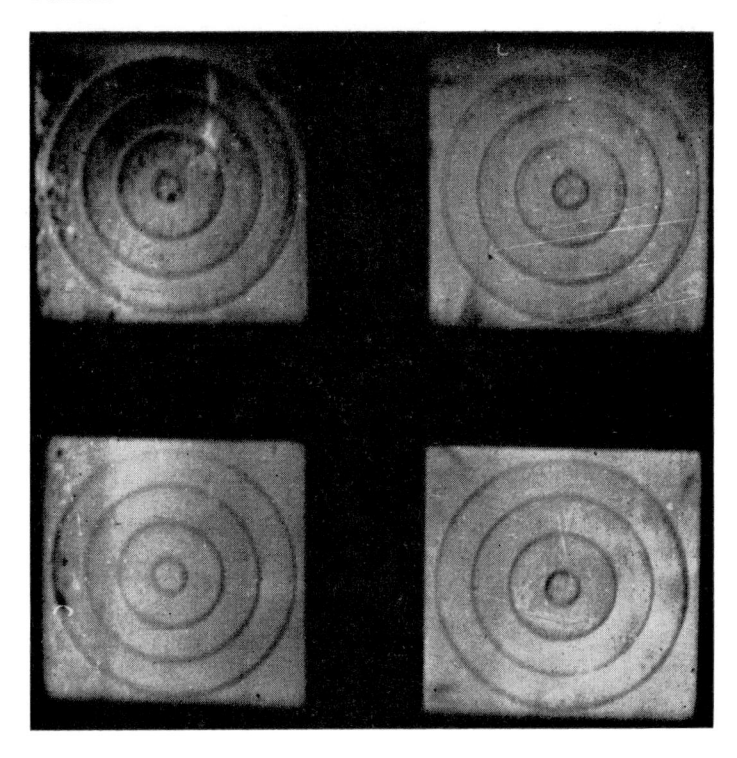

that one had been broken, presumably by some mechanical impact, and another one had been removed and replaced. The remaining panels appeared to be free of any significant cracking or crazing and the discoloration was slight. The light transmission is still adequate. It is fair to note that these panels were replaced by similar acrylic panels soon after this inspection was made.

PHENOLIC AND AMINO RESINS

Phenolic resins were among the first polymeric products to be produced commercially. They are formed by the condensation of a phenol with an aldehyde to form a cross-linked polymer structure. The raw materials most often used are phenol and cresol, which are reacted with formaldehyde or furfural. Amino resins cover a range of resinous polymers produced by the interaction of amines with aldehydes. The two most important are urea formaldehyde and melamine formaldehyde. The melamine formaldehyde resins are superior in their durability properties but are more expensive than both urea and phenolic resins.

Phenolic and amino resins are used in the production of commercial laminated plastics. They may be used with various reinforcements; paper, for example, is used in the manufacture of decorative laminates. Because of its high cost it is more usual to impregnate only the surface layers of laminating material with melamine resins and the base layers with less-expensive phenolic resins. This is the principle applied to most of the commercially available decorative laminates (48, 49).

Thermosetting resins such as phenolic and amino tend to be more weather resistant than thermoplastics for the following reasons (27): *a*, the chemical nature of the cross-linking makes them less susceptible to temperature changes; *b*, plasticisers which tend to make some thermoplastics less stable are not usually used with thermosets; and *c*, the opacity of the thermoset moulded products provides a natural screening against harmful effects of sunlight (19). The resistance to degradation of phenol and amino resins depends to a large extent on the type of filler or reinforcement which is used. In the production of decorative laminates, not only is it important to ensure the stability of the base polymer but the papers should also be colour stable to light and so should the printing inks used (50).

Phenolic laminates are used as curtain wall panels, wall linings, corrugated roofing, and so on. The principal use for the melamine-surfaced laminates is for interior working surfaces but they also have been used out of doors. The mechanical properties of decorative laminates do not deteriorate to an unacceptable degree during weathering. However, discolorations can occur. The extent of the discoloration can be influenced by the particular polymer formulation. A melamine laminating resin used for exterior applications should not contain a plasticicizer (50). Selection of an appropriate polymer formulation for the decorative skin overlay resin can minimize yellowing. The nature and type of paper which is used in a laminate can also influence the durability of the colour of panels. Reports in the literature suggests that yellowish-brown papers are particularly liable to colour change, dependent on the resin system used, while the difference between the red and white papers was relatively insignificant.

As has been found with other polymers, the weathering performance depends very much on the formulation of the particular resins used, the quantity of the reinforcement, the colours employed and on the laminating technique used. Laboratory tests to indicate weathering patterns have been reported in the literature, these indicate that loss of gloss and discoloration were the first manifestations of weathering followed later by micro-crazing of the surface.

Case-histories   The two case-histories summarized below are representative of examples noted in the RAPRA study.

Decorative melamine-faced phenolic paper laminates, installed at the entrance of a south-facing shop in Edinburgh in 1948, were recently inspected and it was noted that only slight loss of gloss had occurred and some yellowing of the green print was noted. These particular laminates have been washed regularly by the shop owner who is still completely satisfied with the product.

A dark-brown phenol formaldehyde/paper laminate was inspected at an industrial site; it has been used as a wall-cladding material. This product has been installed fourteen years ago with a recommendation that it should be repainted after ten years or so. The panels to date have not been painted but have become a little lighter in colour and matt. In general they are acceptable in this environment.

In common with many plastics materials these decorative laminates can sometimes exhibit a very fine crazing pattern on ageing. This effect can be minimized by a careful selection of binder resin.

POLYMER FILMS

Much of the weathering effects occur at the surface of an exposed product. Therefore a considerable protection can be achieved if a weather-resistant surface can be applied. The traditional method used on wood and metals is painting.

Recently some highly stable but expensive polymer films have been produced which can be applied to a sheet substrate during its manufacture. Such films include polyvinyl fluoride (PVF) and acrylics. Each of these has been used to surface many types of plastic and other sheets. Probably the PVF product has been most widely publicized.

Polyvinyl fluoride film (51-3)

PVF is generally available in film form which may be transparent or pigmented. Not only has the film excellent weathering characteristics but also its mechanical properties are sufficient to resist fractures due to impact, flexing, and elongation due to tensile forces. The film is also resistant to damage by corrosive atmosphere and abrasion.

It has been used to surface a variety of materials including metals, cellulosic board, fibreboard, plywood, papers, phenol formaldehyde/paper laminates, acrylics, polystyrenes, PVC, ABS, polyurethane, glass-fibre reinforced polyester laminates, and asbestos. It use on GRP has already been described.

Maintenance procedures are simplified when PVF is at the surface since it is resistant to staining by many solvents, staining agents, and industrial deposits. It has excellent release properties which allows it to be wiped clean, and it can, if necessary, be scrubbed without damage to its surface.

Indications are that PVF has a weather-resistance superior to most plastics. Clearly since it is used as a surface film it is important that it should remain in position during the life of the product and thus care must be taken to ensure that the bond to the substrate is firm and consistent. Claims have been made that pigmented PVF film will have a life of twenty-five years or more. During the RAPRA case-history study no PVF-coated panels were examined.

Acrylic film Acrylics are also available in film form which can be attached to various substrates. Excellent weathering characteristics are imparted to the product.

PVC A number of paints based on PVC are available. These generally have good weathering and chemical resistance although this is not as good as that obtained with exterior grades of unplasticized PVC mouldings.

PVC surfaces are applied to metal substrates and a range of coated steels are available. This product has excellent durability and can be shaped and formed by normal cold-pressing techniques.

Polyurethanes — Polyurethane-based varnishes are well known in the building industry for coating wood and metals. They have also been used to treat surfaces of glass-fibre reinforced polyester laminates—particularly roofing panels and hot-pressed mouldings. The varnishes can be formulated to minimize crazing and yellowing on ageing and generally have excellent durability. Care has to be taken in applying the coating which may require stoving to achieve its best properties.

Other surface coatings — Acrylic, epoxy, alkyd, and phenolic-based paints and varnishes are also available.

GENERAL OBSERVATIONS

The published work on the durability and weathering of plastics contains a considerable amount of detail about the changes in chemical structure and compounding which have been undertaken to achieve improvements in performance. Most of the supporting evidence for improvement has been obtained from small samples exposed in an artificial environment or at tropical sites known to be aggressive. Very little data is available on long-term performance in temperate climates over periods significant to the building industry.

The reluctance of authors to predict the life of materials is understandable, since so many reservations may have to be made regarding the use and location of the product. Predictions which are made are often conservative because care has to be taken to take into account the most rigorous conditions likely to be experienced in this country. Thus since heat and ultra-violet light are the troublesome factors, a panel facing south and inclined towards the sun is going to deteriorate much faster than one facing north. Also the effects of pigments on durability must be known.

Plastics

Clearly the data which have been collected in the past increase confidence in the use of PVC, GRP, decorative laminates, acrylics, PVF, polyurethanes, and other materials used outside.

Also there is a problem in deciding what criteria determine the 'life' of the product. Many plastics are selected for exterior application because they offer the architect a greater opportunity to provide decorative effects, whether they be in shape, colour, transparency, or texture. If this is so then loss of colour, transparency, or gloss may mask the intended effect. If, however, the change in appearance on weathering is known and is consistent, the final appearance will possibly be perfectly acceptable. In one of the RAPRA case-histories a church spire in GRP was seen. This originally had been too green and was not entirely satisfactory, however, after a few years the 'brash' colour had changed—it had become slightly darker—and became completely acceptable. Because of an acquaintance with the changes in appearance of wood, brick, or concrete there may be a tendency to consider 'weathering' as 'maturing'.

In general, the plastics materials which are recommended for exterior use have been formulated and processed to give the best weathering characteristics which can be achieved economically with present-day techniques. Retention of colour and appearance is a dominant requirement. These products also retain their mechanical properties and are not likely to fail unless subjected to excessive loading. Sometimes changes in the environment can cause failure—a good example of this type of difficulty is in waste pipes. These were designed to withstand a certain level of heat loading based on the temperature and volume of water used in washing and washing-up machines. For many years no problem arose, but newer machines employed higher temperatures and larger volumes of water. Under the new conditions failure of the waste pipe by distortion occurred.

Dissatisfaction with other plastics products has been due to a failure to apply very simple cleaning procedures. In some cases this was because the user was unaware of the recommendations of the supplier of the materials.

In most cases examined in a pilot study at RAPRA, plastics building components were seen to be durable, functional and easy to maintain. A reasonable confidence in the expected life-span was gained. There were a number of short-comings which have been described in this chapter, In the main they arose from an incomplete appreciation of the performance requirements of the application, the ageing characteristics of the polymer, and installation procedures. As knowledge of these interacting parameters becomes more general and co-ordinated, the growth of plastics in the building industry for exterior applications will be assured.

REFERENCES

1. RUGGER, G. R. (1965). 'Weathering of plastics', *Conf. on Plastics in Building Structures*, paper 16 (London: The Plastics Institute).
2. SCOTT, K. A. (1966). 'Weathering of plastics: weathering factors and exposure tests: a review', *RAPRA Tech. Rev.*, 34 (Shawbury: Rubber and Plastics Research Assoc.).
3. SEARLE, N. Z., and HIRT, R. C. (1962). 'Bibliography on ultraviolet degradation and stablilisation of plastics'. *SPE Trans.*, **2** (1), 32–54.
4. HIRT, R. C., *et al.* (1961). 'UV degradation of plastics and the uses of protective UV absorbers', ibid., **1** (1), 21–5.
5. WALLDER, V. T. (1950). 'Weathering studies on polyethylene', *Ind. Engng. Chem.*, **42,** 2320–5.
6. HIRT, R. C., and SEARLE, N. Z. (1967). 'Energy characteristics of outdoor and indoor exposure sources and their relation to the weatherability of plastics', *Weatherability of Plastics Materials, Applied Polymer Symposia,* **4,** pp. 61–83.
7. HYBRE, R. (1963). 'Action of infrared and ultraviolet radiation on plastics', *Off. Matieres Plast.*, **10** (103), 397–400, 403–4.
8. WRIGHT, B. (1963). 'The ageing of plastics, Parts I & II; *Plastics,* **28** (311), 111–3.

Plastics

9. HEAP, W. M. (1965). 'Microbiological deterioration of rubbers and plastics', *RAPRA Inf. Cir.* no. 476.
10. PENN, W. S. (1962). *PVC Technology* (London: Maclaren).
11. MATTHAN, J., WIECHERS, M., and SCOTT, K. A. (1968). 'A review of the literature on the ageing and weathering of plastics, 6. PVC', *RAPRA Tech. Rev.*, B. 381.
12. PENN W. S. (1964). *Plastics in building handbook* (London: Maclaren).
13. PLUMB, D. S. (1965) 'A realistic look at plastics in building—rigid PVC', *Mod. Plastics*, **42** (6), 75–7.
14. CROWDER, J. R. (1965) 'Plastics at the Building Research Station', *Rubb. Plast. Age*, **46** (7), 809–10.
15. DE COSTE, J. B., and WALLDEN, U. T. (1955). 'Weathering of PVC', *Ind. Engng. Chem.*, **47** (2). 314–22.
16. WEISFELD, L. B., THACKER, G. A., and NASS, L. I. (1965). 'Photodegradation of rigid PVC', *SPE Journal*, **21** (7), 649–58.
17. LEEPER, H. M., and GOMEZ, I. L. (1966). 'Quality standards for rigid PVC for exterior construction', *Mod. Plastics*, **43** (9), 257–60, 269, 394, 396, 400, 402, 406.
18. DE COSTE, J. B., and HANSEN, R. H. (1962). 'Coloured PVC for outdoor applications', *SPE Journal*, **18** (4), 431–40.
19. ESTEVEZ, J. M. J. (1965). 'Some thoughts on the weathering of plastics', *Trans. and Jl. of P. I.*, **33** (105), 89–94.
20. ANON. (1967). 'Playgrounds—the coming field for reinforced plastics', *Mod. Plastics*, **44** (6), 94–7, 173.
21. BOENIG, H. V. (1964). *Unsaturated polyesters* (Elsevier).
22. GILMAN, L. (1957). 'The resistance of glass fibre reinforced laminates to weathering', *SPE Journal*, **13** (11), 33–8.
23. RUGGER, G. R., *et al.* (1966). 'Fire retardant resins', *Weathering of glass reinforced plastics*, p. 36 (New Jersey: Plastics Tech. Evaluation Centre, Picatinny Arsenal, Dover).
24. SMITH, A. L., and LOWRY, J. R. (1960). 'Long-term durabilities of polyesters in glass-reinforced parts', *Plastics Tech.* **6** (8), 50–4 and 56.
25. SMITH, A. L., and LOWRY, J. R. (1958). 'Unsaturated polyester resins containing methyl methacrylate monomer', *Mod. Plastics.*, **35** (7), 134–42, 200.
26. PENN, W. S. (1966), *GRP Technology* (London: Maclarens).
27. RUGGER, G. R. (1964). 'Weathering properties in plastics', *Mat. in Des. Engng*, **59** (1), 70–84.
28. HIRT, R. C., SEARLE, N. Z., and SCHMITT, R. G. (1961). 'Ultraviolet degradation of plastics and the use of protective ultraviolet absorbers' *SPE Trans.*, **1** (1), 21–5.
29. SEARLE, N. Z., and HIRT, R. C. (1962). 'Bibliography on ultraviolet degradation and stabilisation of plastics', *ibid.*, **2** (1), 32–54.
30. CROWDER, J., and MAJUMDAR, A. J. (1968). 'Weathering of GRP' *Plastics* (Sept.), pp. 1012–3.

31. BRINDLEY, S. H. (1965). 'The weather resistance of glass reinforced polyester sheeting', *British Plastics*, **38** (4), 224–6.
32. WHITEHOUSE, A. A. K., and WILDMAN, D. (1964). 'Surface weathering characteristics of reinforced polyesters', Paper 28 *4th International Reinforced Plastics Conference* (London).
33. *Cellobond Unsaturated Polyester Handbook*, TM12. Plastics Division, BP Chemicals Ltd.
34. SCOTT, K. A. (1959). 'Effects of weathering on colour of polyester glass laminates', *British Plastics*, **32** (1), 112.
35. SONNEBORNE, R. H. (1959). 'Properties of commercial translucent reinforced plastics panels', *14th Ann. Tech. & Man. Conf. SPI*, Section 12-F (Chicago).
36. *PVF Film*. E. I. Du Pont Nemours & Co., Tech. Bulletin.
37. CRENSHAW, J. B., and SMITH, D. (1962). 'An investigation of the optical properties of fibreglass reinforced translucent panels, Part II, An interim report' *17th Ann. Tech. & Man. Conf. SPI*, Section 11-E (Chicago).
38. SCHLAUB, J. A. (1964). 'Surfacing mats in weathering and chemical resistance of laminates', *19th Ann. Tech. & Man. Conf. SPI*, Section 4-A Chicago).
39. GOLDSBERRY, K. L. (1967). Florists review, 20th April.
40. 'A realistic look at plastics in building—acrylic sheet' *Mod. Plastics*, **42** (12), 89–93, 166. (1965).
41. RUGGER, G. R. (1965). 'Weathering of plastics', *The Plastics Institute Conf. on Plastics in Building Structures*, Paper 16 (London).
42. *Perspex acrylic materials properties*, p. 32 (ICI Properties: Welwyn Garden City).
43. RUSSELL, E. W. (1950). 'Crazing of cast PMMA', *Nature*, **165** (4186), 91–6.
44. *Cast and extruded methacrylic sheet for forming operations*, Montecatini Technical Report N.T. Dies no. 120–E.
45. ICI LTD. (1964). 'Antistatic agents for perspex acrylic sheet', *ICI Inf. Service Note*, 1106, p. 2 (Welwyn Garden City: ICI).
46. —— (1965). 'Weathering properties of coloured perspex', *ICI Plast. Divn., Techn. Data PX TD 206-7*, p.2.
47. BUILDING RESEARCH STATION (1966). 'Applications and durability of plastics', *BRS Digest*, no. 69 (Second Series). (London: HMSO).
48. *Building with plastics* (1966). Shell Plastics Advisory Service.
49. BIP CHEMICALS LTD. (1965). *Weathering of exterior melamine laminates*.
50. 'Tough hide for building materials' (1962), *Plastics World*, p. 14.
51. 'PVF film protects laminates' (1963), *Plastics Tech.*, **9**, 56.
52. 'Rubberoid goes into plastics roofing' (1964), *Plastics and Rubber Weekly*.
53. 'Tedlar' (1964), *Jl. R. Inst. Chem.*, **86**, 56.

BIBLIOGRAPHY

General references

BIKALES, N. M. (1967). 'New weathering data', *Plastics Technol.*, **13** (4), 11–13.

BLINNE, F. J. H., and DAY, L. E. (1961). 'Resistance of plastics to outdoor exposure', *US Govt. Res. & Develop. Rept.*, **40**, no. 3.

BURGESS, A. R. (1952). 'Degradation and weathering of plastics', *Chem. & Ind.*, pp. 78–81.

CLARK, J. E., and HARRISON, C. W. (1967). 'Accelerated weathering of polymers: radiation', *Weatherability of Plastic Materials, Applied Polymer Symposia*, **4**, 97–110.

CHOTTINER, J., and BOWDEN, E. B. (1965). 'How plastics resist weathering'. *Mater. Design Engng.* **62** (4), 97–9.

COVES, J. (1967). 'Durability of plastics', *Rev. Plast. Mod.*, no. 137, 869–71.

CUTHRELL, R. E. (1967). 'Environment influenced surface layers in polymers', *Jl. Appl. Polym. Sci.*, **11** (8), 495–507.

ESTEVEZ, J. M. J. (1965). 'Some thoughts on the weathering of plastics', *Plast. Inst. Trans.*, **33**, (105), 89–94.

GRAY, V. E., and WRIGHT, J. R. (1964). 'Pinpointing effects of sunlight on plastics', *Plastic World*, **22**, (12), 14–15.

GRINSFELD, H. (1967). 'Analysis of plastic weathering results', *Weatherability of Plastics Materials, Applied Polymer Symposia*, **4**, 245–62.

GOLDFEIN, S. (1966). 'Prediction of mechanical behaviour of plastics undergoing decomposition from the combined effects of environmental exposure and stress', *Jl. Appl. Polym. Sci.*, **10**, (11), 1737–50.

HAMILTON, C. W., *et al.* (1954). 'Study of outdoor ageing and weathering', *Final Rept. US Dept. of the Army Signal Corps*, Contract No. DA–36–039, SC–15436.

HAUK, J. E. (1965). 'Long term performance of plastics', *Mater. Design Engng*, **62**, (6), 113–33.

HILDEBRAND, C., and MUTH, T. (1967). 'Weathering of plastics in buildings, *Plaste u. Kaut.*, **14** (4), 263–4.

INGLE, G. W. (1964). 'Ageing of plastics in structural applications', *SPE Trans.*, **4** (3), 224–8.

JELLINEK, H. H. H. G. (1967). 'Fundamental degradation processes relevant to outdoor exposure of polymers', *Weatherability of Plastics Materials, Applied Polymer Symposia*, **4**, 41–59.

—— (1965). 'Some aspects of polymer degradation processes', *Special Tech. Publ* 382, 3–13, (Philadelphia: ASTM).

KALLS, D. H. (1963). 'Plastics natural weathering program: three and four years weathering', *US Govt. Res. Rept.* 38, no. 22, p. 93.

KAMAL, M. R., and SAXON, R. (1967) 'Recent developments in the analysis and prediction of the weatherability of plastics', *Weatherability of Plastics Materials, Applied Polymer Symposia*, **4**, 1–28.

KINMONTH, R. A. (1964). 'Weathering of plastics', *SPE Trans.* **4** (3), 229–35.

MARK, H. F., and ATLAS, S. M. (1964). 'Principles of polymer stability', *Stability of Plastics*, Preprints of SPE Conference, pp. 5–13.

MASSEY, P. H. (1966). 'Results of plastic weathering experiments', *Mater. Plast. Elast.*, **32** (3), 317–20.

MATSUDA, T., and KURIHARA, F. (1965). 'The effect of humidity on the ultraviolet oxidation of plastic films', *Chem. High Polymers, Japan*, **22** (243), 429–34.

McNALLY, C. (1963). 'Resistance of plastics to outdoor exposure', *US Govt. Res. Rept.* 38, no. 21.

NEIMAN, M. B. (ed.) (1965). *Ageing and stabilisation of polymers* (New York Consultants Bureau).

—— (1964). 'Prolongation of the life of polymers', *US Joint Publ. Res. Service*, JPRS: 26843 (Washington).

PINNER, S. H. (1966). *Weathering and degradation of plastics* (Manchester: Columbine Press).

PLASTICS INSTITUTE OF AMERICA INC. (1963). 'Ageing of plastics', *Mod. Plastics*, **41** (2), 380–2.

RAPHAEL, T. (1962). 'Predicting service life of plastics', *Plast. Technol.* **8** (10), 26–8.

RUGGER, G. R. et al. (1966). *Weathering of glass reinforced plastics*, (US Dept. of Commerce).

SEARLE, N. Z., and HIRT, R. C. (1962). 'Bibliography on ultraviolet degradation and stabilisation of plastics', *SPE Trans.*, **2** (1), 32–54.

STUART, H. H. (1967). 'Physical causes of ageing in plastics', *Angew, Chem. Int. Ed.*, **6** (10), 844–51.

WINSLOW, F. H., and HAWKINS, W. L. (1967). 'Some weathering characteristics of plastics', *Weatherability of Plastic Materials*, Applied Polymer Symposia, **4**, 29–39.

WOLKOBER, Z. (1967). 'The ageing of plastic systems', *Plaste u. Kaut.*, **14** (3), 141–7.

YUSTEIN, S. E., and WINANS, R. R. (1961). 'Report of the investigations of outdoor weathering of plastics under various climatological conditions', *US Govt. Res. Rept.*, 36, no. 7.

ANON. (1967). 'The score on weatherability', *Mod. Plastics*, **44** (9), 86–91, 162, 164, 166, 170, 175.

Polyvinyl chloride

CHEVASSUS, F., and BROUTELLES, R. de. (1963). 'Influence of Lubricants', *The stabilisation of polyvinyl chloride* (Trans.).

CROWDER, J. R. (1965). 'Plastics at the Building Research Station', *Rubber Plast. Age*, **46** (7), 809–10.

VESCE, V. C. (1959). 'Exposure studies of organic pigments in paint systems', *Official Digest*, **31** (419) (Part 2).

JASCHING, W. (1962). 'Uber Abbau und Stabilisierung von Polyvinylchloride', *Kunststoffe*, **52** (8), 458–63.

SMITH, H. V. (1962). 'Sidste Nyt om PVC—Stabilisatorer', *Plastic*, **12** (4) 134.

LANGSHAW, H. J. M. (1960). 'Weathering of High Polymers', *Plastics*, **25** (267), 40–50.

WARTMAN, L. H. (1955). 'Heat and light degradation of vinyl chloride resin', *Ind. Engng Chem.*, **47** (5), 1013–9.

CLARK, F. G. (1952). 'Accelerated and outdoor weathering of coloured vinyl films', ibid., **44** (11), 2697–709.

HENDRICKS, J. G. (1955). 'Weathering properties of vinyl plastics', *Plast. Technol.*, **1** (2), 81.

GEDDES, W. C. 'The mechanism of PVC degradation', *RAPRA Tech. Rev.*, no. 31.

FREY, H. H. (1963). 'Light and weather resistance of high impact PVC', *Kunststoffe*, **53** (2), 103–10.

FOX, W. V., HENDRICKS, J. G., and RATTI, H. J. (1949). 'Degradation and stabilisation of PVC', *Ind. Engng Chem.*, **41**, 1774–9.

DYSON, G. M., HOVOCKS, J. A., and FENNLEY, A. M. (1961), 'Some aspects of PVC stabilisation and degradation', *Plastics*, **26** (288), 124–8.

HENDRICKS, J. G., and WHITE, E. L. (1951). 'Stabilisation of PVC-type plastics', *Ind. Engng Chem.*, **43** (2), 2335–8.

JACOBSON, V. (1961). 'The influence of lubricants on rigid PVC', *British Plastics*, **34** (6), 328–34.

GRAHAM, P. R., DERBY, J. R., and KATLAFSKY, B. (1958). 'Improved light stability of plasticised PVC', *ACS Div. of Paint, Plast. & Printing Ink Chem.*, **18** (1), Paper 36, 239–55.

WORMALD, G., and SPENGEMAN, W. F. (1952). 'Pigment colours for plasticised vinyl chloride polymers', *Ind. Engng Chem.*, **44** (5), 1104–7.

ANON. (1966). 'Assessing durability of plastic roofs', *Aust. Plast. Rubb. Jl*, **21** (252), 25–6.

TATEVOSYAN, G. O., and KUZNETSOVA, I. B. (1962). 'Weather resistance of film materials', *Soviet Plastics*, no. 3, 38–43.

BONNIE, S. (1957). 'Colorants for plastics', *Plastics Technol.*, **3** (8), 633–8.

MACK, G. P. (1953). 'Stabilisation of polyvinyl chloride', *Mod. Plastics*, **31** (3), 150–4, 218–26.

DEANIN, R. D. (1966). 'Stabilisation of plastics', *SPE Journal*, **22** (9), 13–17.

HOPKINS, R. P. (1960). 'Acrylic polymers for use in rigid and semi-rigid vinyl compounds', ibid., **16** (3), 304.

BAUM, B. (1961). 'Polyvinyl chloride degradation and stabilisation', ibid., **17** (1), 71–6.

GRIEFF, B., and PORTINCELL, G. C. (1963). 'Injection moulding of unplasticised PVC compounds', *British Plastics*, **36** (6), 319–25.

SCARBROUGH, A. L., KELLNER, W. L., and RIZZO, P. W. (1952). 'Role of HCl in PVC degradation', *Mod. Plastics*, **29** (9), 111–22.

DARBY, J. R., and GRAHAM, P. R. (1962). 'Outdoor durability of plasticised polyvinyl chloride', ibid., 39 (5), 148–57, 168.

PERRY, N. L. (1963). 'Stabilisation of rigid PVC for outdoor exposure', ibid., **40** (9).

REINER, G. M. (1966). 'Problems in the colouring of plastics', *RAPRA Translation* 1371.

VINCENTE, L. M. (1966). 'The problem of colour in plastics', ibid., 1365.

ROBIN, M. (1965). 'Stabilisation of PVC using barium compounds', *RAPRA Inf. Circ.* 477.

Glass-fibre reinforced polyester resins

SMITH, A. L., and LOWRY, J. R. (1959). 'Factors affecting durability of glass-reinforced polyester plastics', *Plastics Technol.*, **5**, (6), 42–8, 50, 56.

RAWE, A. W. (1962). 'Environmental behaviour of glass-fibre reinforced plastics', *The Plastics Institute, Trans. and Jl.*, **30** (85), 27–35.

PARVIN, K. (1965). 'The effect of resin elongation at break on the properties of glass-reinforced plastics', ibid., **33** (107), 141.

PARKYN, B. (1959). 'Self extinguishing polyester resins', *British Plastics*, **32** (1), 29.

DEAN, R. T., and MANASIA, J. P. (1955). 'Stabilisation of polyesters to light', *Mod. Plastics*, **32** (6), 131–8.

PUSEY, B. B., and CAREY, R. H. (1955). 'Effects of time, temperature and environment on the mechanical properties of polyester-glass laminates', ibid., **32** (7), 139.

MATHESON, L. A., and BOYER, R. F. (1952). 'Light stability of polystyrene and polyvinylidene chloride', *Ind. Engng Chem.*, **44** (4), 867–73.

BROOKFIELD, K. J., and PICKTHALL, D. (1959). 'The effect of heat and water immersion on the flexural strength of certain glass fibre polyester resin laminates', *Reinforced Plastics*, **4** (4), 13–17.

BOENIG, H. V., and WALKER, N. (1961). 'Shrinkage of glass reinforced polyesters', *Mod. Plastics*, **38** (6), 123.

HIRST, R. C., SEARLE, N. Z., and SCHMITT, R. G. (1960). 'Ultraviolet degradation of polyester resins and the use of protective ultraviolet absorbers', *15th Ann. Tech. & Man. Conf. SPI.* Section 10–A (Chicago).

—— and SCHMITT, R. G. (1959). 'Solarisation studies on polyester resins', *14th Ann. Tech. & Man. Conf. SPI*, Section 12–A (Chicago).

SCHLARB, J. A. (1964). 'Surfacing mats in weathering and chemical resistance of laminates', *19th Ann. Tech. & Man. Conf. SPI*, Section 4–A (Chicago).

SMITH, A. L., and LOWRY, J. R. (1958). 'Some factors influencing the durability of glass-reinforced polyesters', *13th Ann. Tech. & Man. Conf. SPI*, Section 6–B (Chicago).

TORRES, A. F., YOVINO, J., and BOEKER, B. E. (1962). 'A study of the variables affecting the corrosion resistance of premix', *17th Ann. Tech. & Man. Conf. SPI*, Section 5–E (Chicago).

YUSTEIN, S. E., and WINANS, R. R. (1952). *Report of investigation of outdoor weather ageing of plastics under various climatological conditions*. US Department of Commerce.

Properties of polyester coats and laminates, (1965), pp. 141–71. Cellobond Polyester Resins, Technical Manual no. 12 (British Resin Products Ltd.).

Polyester handbook (1969). (Scott Bader & Co. Ltd.).

PLASTICS INSTITUTE (1965). *Conference on plastics in building structure* (London).

CYWINSKI, J. (1960). 'Methyl ethyl ketone peroxide' and 'Benzoyl peroxide', *The role of peroxides in curing polyester resins and their influence on the physical properties of reinforced plastics*, pp. 29, 38 (Novadel Ltd.).

MOSS, W. H., and LANG, A. B. (1964). 'The use of accelerated weathering methods in the development of reinforced plastics products for outdoor use', *4th International Reinforced Plastics Conference* Paper 26 (London).

CROWDER, J. R. (1964). 'The weather behaviour of reinforced plastics sheeting', ibid., Paper 27 (London).

RUGGER, G. R. (1964). 'Fifteen years of weathering results', *SPE Trans.* **4** (3), 236–40.

PARKYN, B., and HULBERT, G. C. (1958). 'Twelve years of reinforced plastics', *1st Reinforced Plastics Conference*, Section J (Brighton).

—— (1954). 'Ancillary materials', *Glass reinforced plastics*, ch. 4, pp. 57–68 (Iliffe & Sons Ltd.).

SONNEBORN, R. H., and BASTONE, A. L. (1960). 'Properties of commerical translucent panels (Part 2)', *15th Ann. Tech. & Man. Conf. SPI*, Section 2–D (Chicago).

HICKS, J. S. (1947). 'Reinforcements for plastics and sandwich structures', *Low-pressure laminating of plastics*, ch. 4, pp. 59–92 (New York: Reinhold).

LAWRENCE, J. R. (1960). 'Introduction and history', *Polyester Resins*, chap. 1, pp. 6, 10 (New York: Reinhold).

KAMAL, M. R., and SAXON, R. (1967). 'Recent developments in the analysis and prediction of the weatherability of plastics', *Weatherability of Plastics Materials, Applied Polymer Symposia.* **4,** 1–28.

BUCKNALL, C. B. (1966). 'Degradation and weathering of polystyrene and styrenated polyester resins', *Weathering and Degradation of Plastics*, Pinner, S. H., (ed.) ch. 6, pp. 81–103 (Columbine Press).

Acrylics

FROLOVA, M. I., EFINOV, L. I., and CHEKMODEEVA, I. V. (1965). 'Ageing of polymethyl methacrylate organic glass under the action of radiation from sun-ray lamps', *Soviet Plastics*, no. 3.

TRUSOVA, K. I., and GUDIMOV, M.M. (1962). 'Ageing of PMMA under the action of atmospheric factors and the stresses encountered in use', ibid., no. 9, 39–40.

VORONOVSKII, N. E., and VOSTRESENKII, V. A. (1967). 'Aging of PMMA under subterranean conditions', *Plast. Massy*, no. 2, 71–3.

LEDBURY, K. J., and STOKE, A. L. (1965). 'Accelerated ageing of plastics. 1. Degradation of thermoplastic materials under dry and humid/hot conditions', *Explosives Res. & Dev. Establishment, Tech. Memo* 18/M/64 (Waltham Abbey: Ministry of Aviation).

JORDON, J. M., MCILORY, R. E., and PEARCE, E. M. (1967). 'Accelerated ageing of PMMA copolymers and homopolymers'. *Weathering of Plastics Materials, Applied Polymer Symposia*, **4,** 205–18.

PEIRSON, O. L. (1965). 'Acrylic plastics in building construction', *Appl. Plast. Reinf. Plast. Rev.* **8** (3), 17–20 and 31.

MARTIN, K. G. (1966). 'Assessing durability of plastic roofs', *Aust. Plast. Rubb. Jl.* **21** (252), 25–6.

'Oroglass Sheet, Physical Properties' (1965). *Oroglass Bulletin*, no. 229 i. (Lennig. Chem. Ltd.)

RAE FARNBOROUGH (1957). 'The effect of tropical exposure on high softening point "perspex"', *Ministry of Supply Technical Note* no. CHEM 1305. (London: HMSO).

FOX, R. B., ISSACS, L. G., and STOKES, S. (1963). 'Photolytic degradation of PMMA', *Jl. Polym. Sci.*, **A1** (3), 1079–86.

Phenolic and amino resins

RUGGER, G. R. (1964). 'Weathering resistance of plastics', *Mat. Des. Engng*, **59** (1), 69–84.

ESTEVEZ, J. M. J. (1965). 'Some thoughts on the weathering of plastics', *The Plastics Institute Trans. Jl.*, **33** (105), 89–94.

'3—Aminoplastic mouldings' (1955). *Reports on plastics in the tropics*. (Ministry of Supply).

'4—Phenolic mouldings' (1956). ibid. (Ministry of Supply).

'8—Asbestos filled phenolic resin' (1958). ibid. (Ministry of Supply).

'9—Phenolic resin bonded paper tube' (1958). ibid. (Ministry of Supply).

'10—Synthetic resin bonded laminated sheet' (1958). ibid. (Ministry of Supply).

LONG, J. K. (1949). 'Effect of exposure', *British Plastics* **21** (246), 619–25.

GREATHOUSE, G. A., and WESEL, C. J. (ed.) (1954). *Deterioration of Materials* (New York: Reinhold).

Polyvinyl fluoride

SIMIEL, V. L., and CURRY, B. A. (1960). 'The properties of polyvinyl fluoride film', *Jl. App. Polymer Sci.*, **4** (10), 62–8.

BEAUDOT, J. E. (1964). 'Film break-through promises protection from weathering', *Canadian Plastics*, p. 44. (December).

Polyurethane surface coating

ROBINSON, E. B., and WATERS, R. B. (1951). 'Urethane oils', *J. Oil and Colour Chem. Assoc.*, **34** (374), 371–3.

SATAS, D. (1963). 'White elastomeric polyurethanes', *The Rubber Age*, **93** (5), 758–60.

'Table IV—Florida exposure for one year, south 45°', *Oil-based One-can Urethane Coatings* (911), p. 6 (Allied Chemical Co.).

KUBITZA, W. (1964). 'Protective finishes on railway rolling stock with new types of light-fast polyurethane coating compositions', *Aust. Plast. Rubb. Jl.*, **19** (229), 23, 26.

MENNICKEN, G. (1966). 'New developments in the field of polyurethane lacquers', *J. Oil and Colour Chem. Assoc.*, **49** (8), 639–47.

WELLS, E. R., et al. (1967). 'Advances in light-stable polyurethanes', *ACS Division of Organic Coatings and Plastics Chemistry*, Papers presented at 153rd meeting in Miami Beach, **27**, no. 1, 263–9.

MENNICKEN, G. (1964). 'Light-fast polyurethane coatings', *Paint Technol.*, **28** (11), 30–2.

ASHTON, H. E. (1967). 'Clear finishes for exterior wood', *Jl. Paint Technol.*, **39** (507), 212–24.

INDEX

N.B. Where the page number is shown in bold type this normally indicates a main discussion of the indexed subject.

Abrasive blasting	65, 83, **102**
Acid washing	61, 78
Acrylonitrile butadiene styrene (ABS)	233, 265
Acrylics (PMMA)	**258**
applications	233, 258
case histories	**260**
decorative	260
effect of heat	260
effect of UV	235
electrostatic charges	259
fabrication	259
films	265
formulation	258
life	258 ff
maintenance and cleaning	**259**
properties	258
reinforced acrylic laminates	**260**
Aggregates	41, 42
calcined flint	64, 88
calcite spar	64
drying shrinkage	43
grading	64, 88
granite	64
limestone	64, 101
silica sand	64
supply	**87**
transparency	58
Air-entraining	44
Alloys (also see aluminium, copper, lead, steel and zinc)	
corrosion resistance	192
heat treatable	195
multiphase	191
non-heat-treatable	194
single phase	191
solution treatment	195
Aluminium and aluminium alloys	**194**
anodizing	196, 200
applications	201
casting	195
cladding	194, 199, 202
compatibility with timber	204
compatibility with plaster	203
compatibility with cement	203
compatibility with steel	202
compatibility with wood preservatives	204
contact with bricks	204
contact with lead	219
contact with steel	202, 203, 225
contact with zinc	209
corrosion resistance	192, 195, **196**, 204
effect of atmospheric exposure	**196**
electrolytic oxidation	196
enamelling	196
fixings for	202
heat treatable	**195**
in aggressive atmospheres	200
interaction with copper	203
maintenance	201
non-heat-treatable	**194**
organic dyes for	196, 200
painting	196, 200
passivity	26, 192
performance on building surfaces	**200**
pitting	199
plastic coated	201
polymeric finishes	196
rainwater goods	203

Aluminium *continued*.
  sacrificial protection 200
  selection of alloys and
    finishes **194**
  solution treatment 195
  super-purity 194, 195, 198
  tensile strength, loss of 198
  weathering characteristics 199
  weathering tests 197
  work-hardening 194
Anodizing 196, 200
Arrises 100
Asbestos cement 8
Atmospheric pollutants 2, 15, 23, **33**, 81
  effect on plastics 235
  effect on wood 140
  effect on zinc 206
  staining of buildings 21, 27

Biological growths 14
  on concrete **62**
  on plastics 235
  on timber 146
Bimetallic corrosion 189, 191, 203
Blockboard 145
Bricks—see Clay products
Bush-hammering 65, **100**

Calcined flint 64, 88
Calcite spar 64
Calcium sulpho-aluminate **124**
Causal factors of deterioration **13** ff
Cavity walls 131
Cements 41, 42
  asbestos cement 8
  cement-polyvinylacetate
    grout 93
  compatibility with
    aluminium 203
  content in concrete 43, 89, 100
  rendering, cracks in 125
  source of soluble salts 113
  sulphate resistant 87, 125
  supply 87
  water-cement ratio 43
  white 59, 81
Cement-polyvinylacetate grout 93
Cladding (see individual materials)
Clay products
  bricks **105**
  black-hearted bricks 116
  boundary walls 109
  brick slips 121
  cavity walls 131
  common bricks 106
  contact with aluminium 204
  contractor's responsibilities **130**
  cycling test 108
  design and construction **126**
  designer's responsibilities **128**
  dimensional changes in **117** ff
  efflorescence **111** ff
  —, removal of 116
  engineering bricks 106, 108
  exposure conditions 129
  frost attack **107**
  handling damage 110
  internal quality bricks 110
  lime bursting 110
  maintenance 132
  moisture movement **119**
  mortar joints 118
  movement joint 121, 123, 129
  pitting 111
  quality control 108
  rendering, cracks in 125
  saturation coefficient 108
  saturation freezing test 108
  soluble salts **112** ff
  spalling 22, 107, 111, 121, 124
  special quality bricks 113, 125
  specification for **127**
  sulphate attack 112, **124**
  thermal expansion,
    coefficient of 118
  thermal movement **118**
  trowelling damage 111
  water absorption 107
  workmanship 127, 130
  zone test 109
Cleaning
  acrylics **259**
  concrete 81, **83**
  GRP **255**
  plastics 268
Climate **28** ff
  macro-(regional)-climate 29, 30
  meso-climate 31
  micro-climate 28 ff
  urban topo-climate 29
Concrete
  abrasive blasting 65, 83, **102**
  accelerated weathering 82
  acid washing 61, 78
  aggregates (also see
    Aggregates) 41, 42, 58

Index

air-entrained	44	—, effect on colour	54
appearance, factors affecting	**49** ff	—, effect on shrinkage	47
biological growth	**62**	—, for abrasive blasted finishes	102
blemishes, definition of and causes	**50** ff	—, for bush-hammered finishes	100
blow-holes	60, 97	—, for white cement concrete	64
board-marked	50, **75,** 86, 88, **98**	modelled façade	**77**
bush-hammering	65, **100**	modulus of elasticity	47
cement content	43, 89, 100	moisture movement	57
chemical resistance	42, **45**	mould oils	**60**
cleaning	81, **83**	patterned surfaces	72, **77**
clear coatings for	**81**	performance, factors affecting	**42**
cohesiveness	64	permeability	**43**
cold weather concreting	**92**	physical irregularities	**52**
colour variations	**53,** 60, 64	placing	**90**
compaction	43, **90,** 94	pore structure	44
compatibility with aluminium	203	precast	41, 78
corrosion of reinforcement	42	prestressed	41, 47
cover to reinforcement	**43,** 100	profiled surfaces	**77**
crazing	**59,** 92	rainwater run-off	
creep	**47**	—, design considerations	69
cube strengths	47	—, effect on weathering	74
deformation	42, **46**	—, modelled façades	80
design	**69**	—, over exposed aggregate	75
durability	**42**	ready-mixed	89
efflorescence	**62**	reinforced	41, 121
elastic modulus	**47**	reinforcement	48, 92, 100
Elastic strain	**47**	release agents	**60**
exposed aggregate	**75**	—, applications	96
—, production of with retardants	60	—, types	**89**
—, production of by acid washing	78	rust staining	**92**
—, effect of silicones on	80	sample panels	90
—, mix design for	65	segregation discoloration	58, 74
—, use of	72	shrinkage, drying	43, **47**
fairfaced	80, 88	shrinkage, reinforced concrete frames	**121**
formwork – see also Formwork	**94** ff	silicones	**80**
frost resistance	42, **43,** 98	spalling	22, 101
gap-grades	**65**	specification for production of high quality finishes	**85** ff
grades, for strength and durability	**44**	structural considerations	**45**
hydration discoloration	54, **57,** 74	sulphate attack	45, **62**
in aggressive atmospheres	54, 64	surface blemishes, causes of	**50**
insulation of	92	surface form	65, **69** ff
joints, construction	46, 86, **91**	thermal movement	**48,** 121
laitance	60, 91	thermal expansion, coefficient of	**48**
lime bloom	**61**	thermal shock	**92**
making good	**93**	tooled	64, **100**
materials for	**87**	toxic washes	**63**
mix design	**89**	variations in colour of	**53** ff
		vibrations	58, 91

Concrete *continued.*
  water-cement ratio   43
  water content   43, 47, 54
  wetting and drying movment 60
  weathering, factors
    affecting   **67**
  white cement concrete   **63,** 81
  workability
  —, effect of entrained air   45
  —, effect on performance   42
  —, of 'making good' mortar   94
  —, tests for   91
  —, white cement concrete   64
Condensation
  in sandwich wall panels   22
  on aluminium   197, 202
  on timber cladding   154
  on zinc   209
Contractor's responsibilities
  brickwork   **130**
  formwork   **94** ff
  handling damage   110
  joints, concrete   46
  quality control   108
  workmanship   23, 127, 130
Copings   72
Copper   **210**
  alloys   210
  applications on buildings   **213**
  artificially patinated   214
  atacamite   211
  brochantite   211
  cladding   211, 213
  compatibility with building
    materials   215
  compatibility with timber   215
  constructional details and
    precautions   **214**
  contact with zinc   209
  corrosion rates   212, 215
  dpc   215
  effect of atmospheric
    exposure   **211**
  fire resistance   216
  fixings   215
  interaction with other
    metals   203, 215
  movement joints   213
  oxide   211
  patina   211
  performance on buildings   215
  protection in the
    atmosphere   193
  protective lacquer for   214

  rainwater goods   215
Corrosion   15, 34
  alloys   192
  aluminium   **194** ff, 204
  anodic and cathodic areas   191
  bimetallic   189, 191, 203
  copper   **211** ff, 215
  electrochemical nature
    of   187, 191
  lead   **216** ff
  of reinforcement   42
  principles of   **186**
  steel   190, **220** ff, **223** ff
  zinc   **205** ff
Crazing
  concrete   **59,** 92
  phenolic and amino resins   264
Creep   **47**
Curtain walls   7, 20, 224
Cycling test   108

Designing
  aluminium   202
  brickwork   **128**
  causal factor of deterioration **17**
  cladding (timber)   **146**
  concrete   **69**
  design elements   19, 23, 28, 31
  detailing   13 ff, 19 ff, 25
  —, clay products   126
  —, concrete   69
  —, metals   193
  external fabric   4
  formwork   55, 56, **95**
  finish to concrete   86
  post construction study   5
  rainwater run-off   27, 65, 115,
                          203
  windows   20
Detailing—see Designing, detailing
Deterioration (also see individual
    materials)   **11** ff
  causal factors   13 ff
  in appearance   12, 28
  in composition   12
  total   **18** ff
Dimensional changes
  bricks   **117**
  concrete   **46**
  timber   142, 149, 159
Durability   10, 21, 27
  concrete   **42,** 44
  paints   26
  plastics   **234**

Index

Efflorescence	
bricks	**111**
caused by bad detailing	117
concrete	**62**
removal of	116
Elastic strain	**47**
Electrochemical series	188, 189
Environment, effect on	
deterioration	16, 21, **28**
Exposed aggregates—see Concrete,	
exposed aggregate	
Extractives	139
reaction with fixings	151
Fertilizers, source of soluble salts	114
Fibreboard	**145**
Fixings	
for aluminium	202
for copper	215
for GRP	257
for lead	218
for timber cladding	**151**
stainless steel	224
zinc	209
Flashings	26
Formwork	
contractor's responsibilities	**94**
definition	86
design of	55, 56, **95**
form-ties	95
glass-fibre reinforced	77, 85
inspection of	**96**
joints	**55**
joints, loss of water	
through	54, 55
non-absorbent	59
plastics	85
plywood	57
prevention of leakage	**97**
release agents	**60, 95, 96**
re-use of	95, 100
rough board finishes	**98**
steel	85
stripping	54, 57
shrinking	**97**
surface finish	95
tolerances in	95
Frost attack	22, 26
bricks	**107**
concrete	42, **43**, 98
Fungal attack	14, **62**, 146, 166
Galvanic cells	**187** ff
caused by deformation	192
concentration cells	192
Galvanizing	205
fixings for aluminium	202
steel	221
Gap grading	65
Glass reinforced plastics (GRP)	**244**
additives	**247**
applications	233, 244
case histories	**255**
cast polyester resin	245
chemistry of polyester resins	245
colour retention	252
composition	
—, effect on gloss retention	**253**
—, effect on fibre pop-out	**253**
—, effect on light	
transmission	**253**
curing	254
discoloration	254
erosion of surface	245, 248
fibre pop-out	**253**
fillers	247
fixing	256, 257
flammability	246, 251
flexural strength	247
formwork	77, 85
gel coat	250
—, effect on gloss retention	**252**
—, effect on light	
transmission	**252**
high transparency panels	251
hot pressing	250
laminates	247, 249
life	254 ff
maintenance and cleaning	**254**
mechanical failure	256
moulding techniques	**248** ff
pigments	247
protection with PVF	254
recovery of gloss and light	
transmission	255
reinforcement types	**247**
spread of flame	252
surfacing tissues	251
—, effect on gloss retention	**252**
—, effect on light	
transmission	**252**
ultra-violet	247
vandalism	256
Gutters, concealed	71
Gypsum	112
Hardboard	145
Hydration discoloration	54, **57**, 74

Joints
  construction (concrete) 86, **91**
  contraction 86
  expansion 86, 119
  failure 8
  formwork **55**
  mortar 118
  movement joints
  —, in concrete 86
  —, in copper 213
  —, in lead 218
  —, in brickwork 121, 123, 129

Laitance 60, 91
Laminates
  GRP 247, 249, 251
  phenol and amino resins 262
Laminboard 145
Lead **216**
  alloys 216
  antimonial 216, 218
  claddings 217
  coating for steel 216
  compatibility with
    building materials 219
  compatibility with timber 219
  constructional details and
    precautions **218**
  contact with aluminium 219
  contact with zinc 209
  corrosion resistance 216
  dpc 219
  effect of atmospheric
    exposure **217**
  fixings for 218
  in aggressive atmospheres 219
  movement joints 218
  paints 217
  performance on building
    surfaces **217**
  protection in the
    atmosphere 193
Lichens 62
Lime bloom **61**
Lime bursting 110
Limestone 64, 101

Maintenance
  acrylics **259**
  aluminium 200
  brickwork 132
  external 10
  GRP **254**

labour force for 9
  costs 3, 7 ff
  plastics 268
  RIBA conference on 6
  zinc 206
Mastics 21, 149, 153
Melamine 262
Metallic soaps 159
Metals (also see particular metals)
  compatibility of 7, 20, 193
  detailing 193
  dissolution in electrolyte 187
  electrochemical series 188
  galvanic series 188
  solution pressure 188
Mix design—see Concrete, mix design
Modulus of elasticity, concrete 47
Moisture content, changes in 141
Mortars
  compatibility with
    aluminium 203
  deterioration 21, 26
  hydraulic lime 113
  making good 93
  mortar joints 118
  sulphate attack **124**
Mosses 62
Mould oils **60**
Mould growth 62, 146
Movement, moisture
  bricks 119
  concrete 57
  timber 141
Movement, thermal
  bricks 118
  concrete **48**, 121
  timber 140

Nails, aluminium 151

Organic dyes 196, 200

Paints
  acrylic 175
  adhesion of 176
  alkyds 175
  aluminium 178
  bituminous 210, 215, 219
  durability 174
  emulsion primer 174
  failure of 10, 21, 31
  finish coats **176**
  galvanized finishes 222

Index

knotting 178
lead 217
maintenance costs 10
  on aluminium 196
  on timber **172**
  on zinc 208
pink primer 174
PVC 265
primers **174** ff
shellac 178
undercoats **175**
weathering **176**
zinc rich 206
Parapets 73
Particleboard **145**
Passivity 26, 192
Patina 211
Performance **1** ff
  studies (also see particular
    materials) 13, 21
Permeability **43**
Phenolic and amino resins **261**
  applications 233, 262
  case histories **263**
  crazing 264
  effect of UV 234
  formulation 263
  life 263
  opacity 262
  stability 262
  tonnage used 232
  weather resistance 263
Photo-oxidation 140
Plants 14
Plaster 115
  compatibility with
    aluminium 203
  effloresence on 115
  gypsum 112
Plastics (also see particular
    plastics) 21, **232**
  additives 233, 235
  applications **232**
  brittleness 234
  cleaning 254, 259, 268
  coated aluminium 201
  coated steel 222
  colour retention 252, 267
  decorative effects 267
  discoloration 234
  durability **234**
  effect of fire 237, 246, 251, 252
  effect of micro-organisms 235
  effect of pollutants 235
  exterior use 267
  factors influencing
    weathering and
    durability **234**
  formulation 233, 236, 245, 258
  formwork 85
  heat distortion 235, 266
  low temperatures 235
  maintenance and cleaning 268
  moulding 232, 248
  stabilization 233, 235
  testing 266
  thermoplastic 232, 235
  thermosetting 232
  tonnage used 232
  ultra-violet 23, 234, 266
Plywood—see Timber, plywood
Polyester resins **244**
Polymer films **264**
  acrylic **265**
  polyurethane varnishes **260**
  PVC **265**
  PVF 181, 233, **264**
Polymethyl methacrylate (see
  also Acrylics) **258**
Polyurethane
  foamed 97
  varnishes **266**
Precast concrete 41, 78
Preservatives—see Timber,
  preservatives
Prestressed concrete 41, 47
Putty 21
PVC **236**
  additives 233
  applications **233**, 236, 239
  —, cladding 236
  —, direction signs 242
  —, floor coverings 240
  —, rainwater goods 232, 239, 240
  —, road signs 239
  —, roof lights 243
  —, surface coatings 240
  —, waste systems 240
  —, water services 239
  —, window frames 240
  —, wire fencing 239
  adjustment of properties 233
  case histories **239**
  coated steel 240
  coatings 265, 266
  discoloration 244
  durability 238

PVC *continued.*
  effect of temperature   236, 243, 244
  effect of UV   234
  extrusion   239
  fillers   237
  films   264
  fire resistance   237
  formulation   236
  life   237 ff
  manufacture and processing   238, 239
  optical properties   238
  paint   265
  stabilization   236
  tonnage used   233
  UB absorbers   237
PVF
  effect of UV   235
  films   **264**
  finishes for timber   181
  life   265

Quality control   108

Rainwater run-off   27
  brickwork   115
  concrete—see Concrete, rainwater run-off
  metals   203
Rainfall source diagram   **31**
Rainwater goods
  aluminium   203
  copper   215
  plastics   232, 239, 240
Reinforcement   47
  corrosion   42
  cover to   **43,** 100
  glass   247
Release agents—see Concrete, release agents
Rendering, cracks in   125
Roof panels (GRP)   251, 255, 256
Roofs, flat   22, 24
  aluminium   201, 202
  copper   211
  mainteance costs   7
  zinc   209
Roofs, pitched   26
  aluminium   201
  zinc   209
Rusting   **190,** 220

Sacrificial protection   200, 205

Saturation coefficient   108
Saturation freezing test   108
Scaffolding   93
Sea   114
Sheradizing   205
Shrinkage, drying
  concrete   43, **47**
  reinforced concrete frames   **121**
Silica sand   64
Silicones   80
Site observations   **19**
Soffits   93
Soluble salts   **112** ff
Spalling
  bricks   22, 107, 111, 121, 124
  concrete   22, 53, 101
Spandrel beams   77
Specifications
  for brickwork   **127**
  for production of high quality concrete finishes   **85** ff
Steel
  carbon   **220**
  contact with aluminium   202, 203, 225
  corrosion rates   220, 222
  effect of atmospheric exposure   **220**
  finishes to   221
  galvanizing   221
  general properties   **221**
  lead coated   216
  low alloy   **221,** 223
  painting of galvanized finishes   223
  performance of galvanized coatings   222
  plastic coated   222
  PVC coated   240
  rusting   **190**
  rust staining   221
  slow rusting   220
Steel, stainless
  austenitic   224
  bimetallic corrosion   225
  compatibility with aluminium   202, 225
  constructional precautions   **225**
  corrosion resistance   224
  cladding   224
  curtain walling   224
  effect of atmospheric exposure   **224**
  ferritic   224

Index

finishes for	**225**
fixings	224
general properties	**223**
in aggressive atmospheres	224
low carbon	226
optical effects	225
passivity	226
uses	**223**
welding	226
Sulphate attack	
bricks	112, **124**
concrete	45, **62**
mortars	124
sulphate resistant cement	124
Surface form	**23** ff, **69** ff
Thermal expansion, coefficients of	43, 118
Thermoplastics	232, 235
Thermosetting plastics	232
Timber	
accelerated weathering tests	179
African mahogany	151
afrormosia	140
afzelia	151
agba	178
beech	162
biological growth	166
blockboard	145
checks	143
chemical and physical nature	**136**
cladding	**146** ff
—, condensation in	154
—, fixings	**151**
—, maintenance	150
—, mastic bedding	149, 153
—, plywood	**153**
—, profiles	150
—, timbers for	151
cleaning	171
colour changes	**140**
compatibility with aluminium	204
compatibility with copper	215
compatibility with lead	219
compatibility with zinc	210
dimensional movement	142, 148, 149, 159, 181
Douglas fir	151
durability	**147**
earlywood-latewood zones	153, 159
effect of atmospheric pollutants	140
effect of UV	166
European redwood	148, 151
factory finishing	**180**
fibreboard	**145**
fibre saturation point	141
film-forming finishes	159, 181
finish coats	**175**
finishes on exterior timber	**156** ff
fungi	166
flatsawn	137, 138
growth rings	138, 139
guarea	151
gurgan	178
hardwoods	136
integral finishes	**180**
iroko	140, 151
keruing	151, 178
laminboard	145
macrostructure of wood	137
meranti	151
metallic soaps	159
microstructure of wood	137
moisture content	147, 148
—, effect on paint	176, 177
nails, aluminium	151
natural exterior finishes	**157**
oak	162, 204, 210
oil-tempered hardboard	145
opaque coatings	**172**
particleboard	**145**
penetrating natural finishes	**167**
—, cleaning	171
—, Madison formula	167
—, paraffin wax	167
—, silicones	169
—, teak oils	167
—, water repellent preservative stains	169
plywood	**143**
plywood formwork	57
plywood cladding	**153**, 155, 170
preservatives	**147**
—, compatibility with aluminium	204
—, effect on paints	178
—, hardwoods	148
—, organic solvents	148
—, softwoods	148
—, water repellent additives	179
primers	**174**
quarter sawn	137
sapele	151

Timber *continued*.		Water–cement ratio	43
softwoods	136	Water repellancy	80
synthetic resinous clear		Weathering tests	
finishes	**164**	accelerated weathering	82
—, edge failure	**165**	aluminium	197
—, moisture curing		bricks	108
polyurethane	164	concrete	82
—, oil modified		copper	212
polyurethane	165	lead	217
—, polyurethane	164	steel	220
—, strippers, use of	165	zinc	206
undercoats	**175**	Weather reports	31
utile	151	White rusting	208
varnishes	**157**	Wind	13, 74
—, blistering	163	Windows	20
—, denibbing	163	Winter concreting	**92**
—, drying oil	158	Wire ties	**93**
—, fillers, use of	161	Wood—see Timber	
—, film failure	160	Woodchecks	143
—, maintenance	166	Workability—see Concrete, workability	
—, penetration	161		
—, resins	157	Work hardening	194
—, viscosity	160	Workmanship	**17,** 127, 130, **180**
veneers	143		
waney-edge boarding	149	Zinc	**205**
weathering	**141**	alloys	205, 206
western red cedar	151, 152, 158,	atmospheric exposure	206
	162, 165, 168, 204, 210, 215,	cladding	208
	219	compatibility with building materials	210
workmanship	180		
Toxic washes	**63**	compatibility with timber	210
Trowelling damage	111	compatibility with wood preservatives	210
Ultra violet radiation	13, 27	contact with aluminium	209
effect on acrylics	235, 258	contact with copper	209
effect on GRP	235, 247	contact with lead	209
effect on phenol and amino		corrosion processes	206
resins	235	corrosion rate	206
effect on plastics	235, 236	effects of atmospheric	
effect on PVC	235, 237	exposure	**206**
effect on timber	142	effects of condensation	209
User, wear by	3, 16	fixings for	209
		galvanizing	202, 205, 209, 221
Vapour barrier	24, 154, 202, 209	maintanance of	206
Varnishes—see Timber, varnishes		painting of	208
Vibration	58, 91	performance on building surfaces	193, **208**
Walls		sacrificial protection	205
boundary	109	sheradizing	205
cavity	131	white rusting	208
curtain	7, 20, 224	zinc-rich paints	206
load bearing	20	zinc spraying	205
sandwich wall panels	22, 24	Zinc Gauge	**226**
Waney-edge boarding	149	Zone test	109